recent advances in phytochemistry

volume 14

The Resource Potential in Phytochemistry

RECENT ADVANCES IN PHYTOCHEMISTRY

Proceedings of the Phytochemical Society of North America

Recent Volumes in the Series

A Continuation Order Plan is available for this series. A continuation order will bring delivery
of each new volume immediately upon publication. Volumes are billed only upon actual ship-
ment. For further information please contact the publisher.

recent advances in phytochemistry

volume 14

The Resource Potential in Phytochemistry

Edited by
Tony Swain
Boston University
Boston, Massachusetts

and
Robert Kleiman
Northern Regional Research Center, USDA
Peoria, Illinois

PLENUM PRESS • NEW YORK AND LONDON

Library of Congress Cataloging in Publication Data

Phytochemical Society of North America.
  The resource potential in phytochemistry.

  (Recent advances phytochemistry; v. 14)
  "Proceedings of the 1979 annual meeting of the Phytochemical Society of North
America, held at Northern Illinois University, DeKalb, Illinois, August 12—15, 1979."
  Includes bibliographies and index.
    1. Botanical chemistry—Congresses. 2. Plant products—Congresses. I. Swain, T.
II. Kleiman, Robert. III. Title. IV. Series. [DNLM: 1. Chemistry, Organic. 2. Plants.
WL RE105Y v. 14/QK861 R434]
QK 861.R38 vol. 14   581.19'2s [581.6'1]   80-23398
ISBN 0-306-40572-5

Proceedings of the 1979 Annual Meeting of the Phytochemical
Society of North America, held at Northern Illinois University,
DeKalb, Illinois, August 12 — 15, 1979.

© 1980 Plenum Press, New York
A Division of Plenum Publishing Corporation
227 West 17th Street, New York, N.Y. 10011

To the memory of
Theodore A. Geissman
1908 — 1978

Theodore A. Geissman
1908-1978

We will miss Ted Geissman. He was a charter member of
our society, which started as the Plant Phenolics Group of
North America. However, his interests were not limited to
just plant phenolics, but to the chemistry of all organic
compounds in living organisms. Our society followed this
course also by expanding its interests to all natural products of plant cells, and renamed itself as The Phytochemical Society of North America. The recent upsurge of
natural products chemistry has a firm basis due to the
pioneering work of Ted Geissman in the study of flavonoids,
terpenoids and alkaloids. Over 200 publications resulted.

Ted was born in Chicago in 1908, and obtained a B.S.
degree in Chemical Engineering in 1930 from the University
of Wisconsin. After 4 years as a chemist with Standard Oil
of Indiana, he returned to academic work and obtained his
Ph.D. in Organic Chemistry in 1939 from the University of
Minnesota. His first natural product chemistry, done
during the next two years as a post-doc at the University
of Illinois, was concerned with gossypol, a dimer of a phenolic sesquiterpene. His research in natural products was
expanded upon his arrival at the Chemistry Department of
U.C.L.A., where he stayed until his retirement in 1974.
The source of his laboratory materials and inspiration was
the nearby desert, and he became a self-educated, highly
competent taxonomist of the local desert flora.

The second World War interrupted this research. In
1943 he became the Associate Director of the Central
Engineering Laboratory of the Office of Naval Research at
the University of Pennsylvania where he worked on a project
concerned with the use of cobalt compounds as a source of
pure oxygen on shipboard. A tragic accident severely
damaged his liver, but he recovered after a long recupera-

tive period to continue effective research and teaching at
U.C.L.A. in 1946. He ultimately died in 1978 of cancer of
the liver.

His popular text, Principles of Organic Chemistry
(Freeman), appeared in 1959, and the 3rd edition with T. L.
Jacobs in 1977. He edited the first monograph on the
Chemistry of Flavonoids, published in 1962 (Pergamon).
This was a major stepping stone in the literature of plant
natural products and is still a useful source of
information. In 1969, he co-authored, with D. H. G. Crout,
a text on the Organic Chemistry of Secondary Plant
Metabolism (Freeman-Cooper).

Many members of our society enjoyed graduate work,
postdoctorals, or visits in his laboratory: Cornelius
Steelink and Anthony C. Waiss, Ph.D. students in 1956 and
1962 respectively; post-docs and research fellows included
Tony Swain ('50), Michael H. Benn, Gretchen Seikel and
Helen Stafford in the 60's. His laboratory was generally
international in its composition. Many post-docs and
research associates came from numerous countries,[1]
including Australia, Brazil, France, Great Britain, Japan,
Netherlands, New Zealand. He traveled to Japan, New
Caledonia, Australia, India, and Taiwan to give symposia.

Ted was also involved in formal teaching at U.C.L.A.,
both in the beginning chemistry course and in graduate
seminars on Natural Products. He trained 50-60 graduate
students at U.C.L.A.; the first received his Ph.D. in 1943,
the last in 1971.

Besides travel to give numerous lectures and for con-
sulting purposes, Ted received major honors such as two
Guggenheim Fellowships (in 1949 and 1964), and was a
Fullbright Research Scholar in 1957. He was on the
Editorial Board of the Journal of Organic Chemistry and of
Organic Procedures.

Ted possessed a wry sense of humor -- well documented
by his reputed comment about the accompanying photograph --
"This rather brooding picture is the only one I find. I
don't know why people have to grin when they are
photographed."[2]

Helen A. Stafford[3]

1. One of them, Edward Leete, has written <u>An Appreciation</u>
    in <u>Phytochemistry</u> <u>18</u>:1259-61.

2. Used also in the AMSOC Courier (Freeman, Cooper &
    Company) No. 3-2000-0376, along with the comment by Ted
    Geissman.

3. With many thanks to Dan Atkinson and the Chemistry
    Department of U.C.L.A. for the photograph and
    bibliographical information.

Preface

This volume of Recent Advances in Phytochemistry is the Proceedings of the 1979 Annual Meeting of the Phytochemical Society of North America held August 12-15 at Northern Illinois University, DeKalb. It contains a series of exciting chapters which start with the potential use of plant products as fuels and medicinals, their possible effects in carcinogenesis and use in steroidal hormone synthesis. The volume continues with a series of chapters which examine the importance of plant constituents in the breeding and selection of corn, cruciferous vegetables, soybeans and citrus fruits. All the contributions illustrate the wide importance of research which improves the health and the economic and social well being of mankind.

The authors are to be congratulated on their lucid exposition of the progress of research in their subject area and for their patience while this book was being produced.

The members of the Phytochemical Society of North America can feel proud of having another of their excellent symposia series in print. It is fitting, therefore, that this volume is dedicated to one of the founder members of the Society, Ted Geissman, who has inspired so many of us with his wisdom, teaching and wonderful support of all our endeavours. He was a giant among phytochemists and is sorely missed by all who knew him.

Finally no meeting can be successful without good organization and we would like to thank, on behalf of the Society, Prof. David M. Piatak of the Northern Illinois University and all members of the Departments of Chemistry and Botany there, who helped to make the meeting a great success. We also wish to thank the DeKalb AG Research Foundation for generous financial support.

Contents

Chapter One

BOTANOCHEMICALS*

R. A. BUCHANAN, F. H. OTEY, AND M. O. BAGBY

Northern Regional Research Center
U.S. Department of Agriculture
Peoria, Illinois 61604

INTRODUCTION

Energy-rich plant products useful as substitutes,
supplements or complements of petrochemicals are now
generally called botanochemicals (a term coined by the USDA
public information officer, Dean Mayberry). Botanochemicals include substances which can be extracted
directly from plants (primary botanochemicals) and those
made by conversion of insoluble saccharides or lignocellu-
loses (secondary or derived botanochemicals). Primary
botanochemicals include products such as naval stores (pine
chemicals), tall-oil products (paper-pulping byproducts),
oils for industrial uses, waxes, tannins, rubber, and gutta
percha. Traditional secondary botanochemicals include

*   The mention of firm names or trade products does not
imply that they are endorsed or recommended by the U.S.
Department of Agriculture over other firms or similar products not mentioned.

furfural and ethanol. Potential new botanochemicals
include latex- or whole-plant oils, guayule rubber, grass
gutta, polyphenols, specialty seed oils, and such potential
lignocellulose conversion products as methane, fuel
alcohol, and various other fermentation-produced fuels and
chemical intermediates.

Now that the price of petrochemicals has increased
faster than the price of equivalent agricultural and
forestry products, the latter have become more attractive.
Various major U.S. petrochemical companies have recently
announced that they are going to make substantial invest-
ments in new technologies for producing pine chemicals,
guayule rubber, fuel alcohol from cellulosic residues,
whole-plant oils, and other botanochemicals. Thus, the
petrochemical industry appears already to have begun a
shift toward a renewable resource base. Undue delay or
failure to make this shift could eventually result in a
"Kondratiev wave" economic recession.[7]

It is especially attractive from both social and economic viewpoints to consider adaptive agricultural systems[12]
for integrated production of primary and secondary
botanochemicals, protein feeds, and fibers. Such systems
have been designated "multi-use botanochemical systems"; a
scenario was presented for their introduction into U.S.
agriculture[4] and economic and feasibility assessments were
made.[5] Such systems can potentially make farm production
of fuels and industrial feedstocks practical without
necessarily decreasing the capacity for food production.

This review surveys and discusses in a cursive way
botanochemical products from multi-use systems. Although
integrated production schemes may be applicable to specialty crops for oilseeds, essential oils, botanicals, plant
fibers, and vegetable dyestuffs, the discussion here rela-
tes to crops designed primarily for extraction of energy-
rich basic feedstocks, that is, oils and hydrocarbons.

BOTANOCHEMICAL PRODUCTION

Systems

Multi-use botanochemical systems offer major advantages

over other biomass utilization schemes (Figure 1). Greater
economy and higher efficiency can be achieved by codevelop-
ment of new crops and new handling, processing, and market-
ing systems than can be achieved just by making better use
of residues from conventional crops. Expensive chemical
conversion of plant products can be minimized by developing
crops specially adapted to the system and capable of
directly producing valuable energy-rich feedstocks. Leafy
residues remaining after extraction of the energy-rich
materials contain most of the plant protein and carbohydrates for feeds and foods, whereas woody stem residues
are especially attractive as lignocellulose sources for
conversion to fuel alcohol. Thus, multi-use botanochemical
crop systems can make possible the agricultural production
of fuels and industrial feedstocks without necessarily
decreasing food production.

Figure 1 portrays only one optional system; crops and
processing can be varied to provide a range of primary
products. In secondary processing of woody residues, other
fermentation products can be produced besides, or in addition to, ethanol. Also, instead of saccharification
followed by fermentation, alternative processing can provide such products as solid fuel pellets, fiber and board
products, cattle feeds, furfural, process gas, methanol,
methane, or even liquid hydrocarbon fuels.

Crops. The U.S. Congress has already selected guayule
(Parthenium argentatum) for development as the crop for
domestic production of natural rubber.[20]

A screening program has so far resulted in identification of more than 40 plant species having some potential as
new botanochemical crops.[2-4] Plant scientists and agrono-
mists are now studying genetics, reproduction, and propaga-
tion of some of these candidates to select species for crop
development. Assuming development of appropriate germplasm
and cultural protocol, specifications for practical botanochemical crops have been proposed.[5] Table 1 compares the
proposed crop specifications with typical compositions and
estimated yields of wild plants, but does not imply that
the new crops will necessarily be developed from the species used for comparison.

In Table 1, botanochemical crops are broadly classified

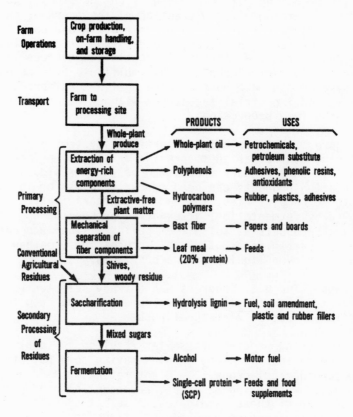

Fig. 1.   Multi-use botanochemical systems.

TABLE 1
Comparison of Botanochemical Crop Specifications with Estimated Yields and Typicals Compositions of Wild Species[a]

Component	Herbaceous perennial oil crop[b] Asclepias syriaca		New crop specification		Woody perennial oil-polyphenol crop[c] Rhus glabra		New crop specification		Perennial grass gutta crop[d] Elymus canadensis		New crop specification	
	Composi-tion, %	Yield, kg/ha/yr	Composi-tion, %	Yield, kg/ha/yr	Composi-tion, %	Yield, kg/ha/yr	Composi-tion, %	Yield, kg/ha/yr	Composi-tion, %	Yield, kg/ha/hr	Composi-tion, %	Yield, kg/ha/yr
Total dry matter	100	12000	100	18000	100	8900	100	11500	100	11200	100	12500
Crude protein[e]	(11.1)	(1332)	(9)	(1620)	(6.7)	(596)	(6)	(690)	(17.2)	(1926)	(10)	(1250)
Gutta	---	---	---	---	---	---	---	---	1.9	213	12	1500
Whole-plant oil	7.6[f]	912	14[f]	2520	5.7	507	10	1150	2.8	314	8	1000
Polyphenol	7.2	864	7	1260	18.8	1673	18	2070	6.6	739	7	875
Leaf meal (20% protein)[e]	16.0	1920	32	5760	10.0	890	10	1150	---	---	---	---
Bast fiber	11.0	1320	6	1080	---	---	---	---	---	---	---	---
Residue[e]	58.2	6984	41	7830	65.5	5830	62	7130	88.7	9934	73	9125

a Yields and compositions are given on a dry weight basis.

b The new crop specification based on harvesting two crops of leafy plant per season. The Asclepias syriaca comparative data is for mature plants harvested in early September, dry yield estimated from individual plant weights and extrapolating to 107,600 plants per hectare.

c Woody plants handled as a coppice crop; harvested by cutting at near ground level, on a 2-year rotation. Dry yield for Rhus glabra estimated.

d Dry matter yields estimated on the basis of one cutting of mature plants per season. The Elymus canadensis data is from an unreplicated small cultivated plot harvested before full maturity.

e Crude protein is not extracted from the plant matter.

f This whole-plant oil contains about 20% low-molecular-weight natural rubber as a hydrocarbon component.

into three types, but actually the primary products will
differ with species. The detailed composition of whole-
plant oils and polyphenols is highly species-dependent.
However, petroleum refinery processes may be insensitive
enough to handle whole-plant oils without regard to crop
species.[8,19] Thus, crop species may be important if chemical intermediates are being produced but noncritical for
production of fuels, solvents, monomers, carbon black, and
other basic chemicals.

Whole-Plant Oils. Whole-plant oils can potentially
become major industrial feedstocks. They are like petroleum and the current commercial plant-derived feedstocks,
naval stores, and tall oil, in that their composition is
very complex. They are much more complex than industrial
vegetable oils consisting mainly of triglycerides from seed
or other plant storage organs. Whole-plant oils generally
contain an entire spectrum of polar to nonpolar lipids
(Figure 2). Their detailed composition depends not only on
the plant species but also on the maturity of the plant and
the method of extraction.[3]

Extraction and partitioning procedures can be varied to
produce, from a given plant sample, oils containing the
different proportions of polar and nonpolar components
desired for different applications. For example, many
temperate zone plants contain low-molecular-weight
rubber[17] that may either be isolated and used as a
polymer or be left as a hydrocarbon component of the oil.

Only a few whole-plant oils have been characterized in
detail, but lipid classes have been estimated by thin-layer
chromatography (TLC) for a defined oil-fraction from
several productive species (Table 2). A distinctive
feature of most whole-plant oils is their high content of
waxy, nonglyceride esters that are less polar than
triglycerides. Frequently, there are two distinct classes
of nonglyceride esters, probably aliphatic wax esters and
triterpenol esters (Figure 2 and Table 2).

Thus, whole-plant oils are valuable mixtures from which
a wide variety of chemical intermediates including sterols,
long-chain alcohols, rosin and fatty acids, esters, waxes,
terpenes, and other hydrocarbons could be obtained.
However, the cost of separation is likely to be high.

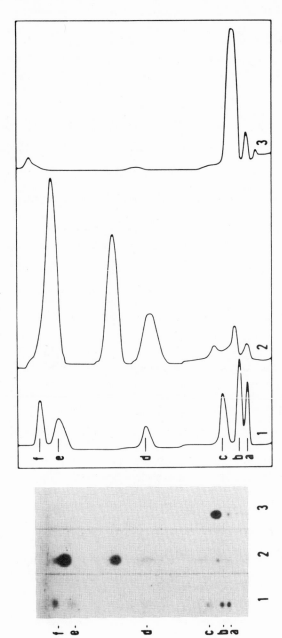

Fig. 2 Thin-layer chromatography of a representative whole-plant oil. Lane 1: mixed standard containing (a) sitosterol, Rf 0.09; (b) oleyl alcohol, Rf 0.12; (c) oleic acid, Rf 0.18; (d) triolein, Rf 0.47; (e) oleyl laurate, Rf 0.80; and (f) squalene, Rf 0.87. Lane 2: Cirsium discolor unsaponifiable matter. The densitometer traces are for the corresponding TLC Lanes with migration distance expanded 1/25 times.

TABLE 2
Lipid Classes in Whole-Plant Oils, Estimated by Thin-Layer Chromatography

Plant source	Harvest date	Oil content, %	Oil Composition					
			Sterols, %	Other free alcohols, %	Free acids, %	Triglycerides, %	Nonglyceride esters, %	Hydrocarbons, %
Ambrosia trifida	9/20/77	4.21	11	4	7	68	7	3
Asclepias incarnata	7/28/77	2.70	9	15	8	13	48a	7
Asclepias syriaca	7/27/77	4.46	5	11	5	trace	72a	7
Cacalia atriplicifolia	8/3/77	2.99	10	24b	7	10	43	6
Campanula americana	9/8/77	6.51	13	5	10	58	10	4
Cirsium discolor	10/5/77	5.66	2	4	4	21	67a	2
Eupatorium altissima	9/20/77	6.08	6	5	7	36	44a	2
Euphorbia dentata	8/25/77	4.13	6	6	5	42	36a	5
Euphorbia Lathyrus	3/25/77	9.21	3	20	18	3c	49	7
Parthenium argentatum	5/13/77	4.04	10	7	5	23	31a	24
Rhus glabra	8/18/77	5.10	12	11	19	13	39	6
Sassafras albidium	9/8/77	2.26	8	46b	--	2	28	16
Sonchus arvensis	6/10/77	4.63	5	19	5	5	60a	6
Vernonia altissima	5/23/77	2.62	13	12	2	4	68a	1

a Prominent nonglyceride ester spot at RF 0.6 in addition to the one at RF 0.8.
b Prominent spot taken as free alcohol at RF 0.23.
c Small spot at RF 0.46 is unsaponifiable.

Accordingly, a likely strategy is to employ crude separations to obtain marketable fractions for various end-uses
(Figure 3). Alternatively, the whole-plant oil can be used
as a feedstock for a petroleum refinery by employing
appropriate process modifications.[8,19] As indicated in
Figure 3, whole-plant oils would be unusually versatile
feedstocks for production of the entire range of petrochemicals as well as tall oil, naval store, and inedible
fat products.

POLYPHENOLS

Although "polyphenols" is actually a generic term
referring to a large complex of phytochemicals with
hydroxy-substituted aromatic rings, we customarily use the
term for crude plant extractives that are at least
sparingly soluble in acetone and freely soluble in 87.5%
aqueous ethanol (Figure 4). If anhydrous methanol or ethanol is used as the extraction solvent, larger yields of
polyphenols of different composition are obtained as
illustrated by Asclepias syriaca (Figure 5). All these
crude products contain, in addition to actual polyphenols,
a wide variety of complex lipids and other substances. For
plants of high tannin content the polyphenol fraction might
equally well be called a tannin extract (note Rhus glabra
in Table 1). Acetone-extracted polyphenol fractions contain 52-60% carbon and have higher heating values than
methanol but considerably lower than whole-plant oils.

Commercial tannins and bark extractives from tree
species used for lumber and pulpwood have been studied with
a view toward expanded industrial utilization[9,15] (Figure
6). Tannin-based plywood adhesives and water-treatment
compounds are commercial products in South Africa. A
polyphenol fraction from creosote bush is used as antioxidant for guayule rubber in Mexico. About 30 million kg/yr
of vegetable tanning materials is imported into the U.S.
However, most of the uses for polyphenols indicated in
Figure 6 are only potential.

If large volumes of low-cost plant polyphenols became
available, they could become very important chemical
feedstocks. However, polyphenol fractions that are likely
products from botanochemical processing have not been

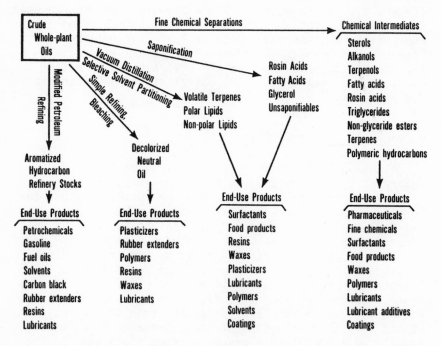

Fig. 3 Strategies for industrial utilization of whole-plant oils

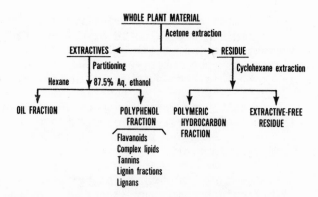

Fig. 4 Separation of crude polyphenol fraction from plant materials

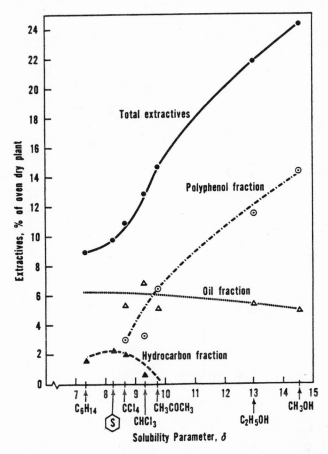

Fig. 5   Yield of polyphenol fraction as a function of solvent
         type.   Exhaustive Soxhlet extraction of <u>Ascepias
         syriaca</u>.

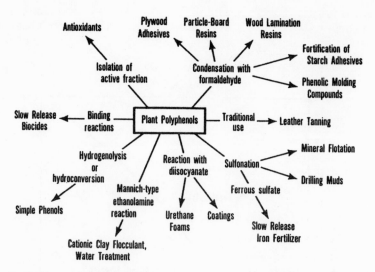

Fig. 6 Routes to industrial utilization of crude plant polyphenols

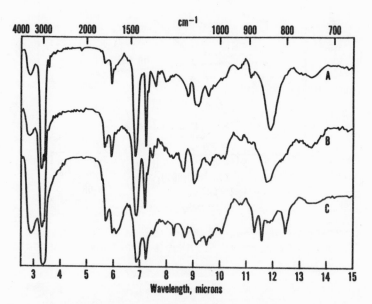

Fig. 7 Infrared spectra. A--natural rubber from Monarda didyma. B--amorphous gutta from Elymus canadensis. C--crystalline gutta from Elymus canadensis

characterized. Their composition is species dependent so
that product and market development can only begin after
the new crop species have been selected. Possibly, conversion to simple phenols would be less sensitive to feedstock
variations than alternative routes to utilization (Figure
6).

HYDROCARBON POLYMERS

cis-1,4-Polyisoprene (natural rubber, NR) is the most
common hydrocarbon polymer found in green plants. Since
the U.S. is now committed to producing NR for polymer uses
from guayule, other crops are less likely to be developed
specifically for high-molecular-weight rubber.

Most of the plant species identified to produce NR contain less than 1%, dry whole-plant basis, of polymer that
is too low in molecular weight[17] for conventional mixing
and processing (Table 3). Low-molecular-weight natural
rubbers would be of interest as a plasticizing additive
(processing aid) to rubber mixes, for liquid rubber processing methods, for making cements (adhesives), and if low
enough in cost as hydrocarbon feedstocks.

It would not be necessary to separate low-molecular-
weight rubber from whole-plant oils intended for use as a
petroleum refinery feedstock. Thus, perhaps the plant
breeder's goal should be to develop crop varieties high in
total content of oil plus low-molecular-weight polyisoprenes.

A few species of grasses may be potential sources of
gutta (trans-1,4-polyisoprene[6]) (Table 4). Known species
contain less than 2% gutta, dry whole-plant basis.
Although the molecular weights of grass guttas are lower
than those of most commercially important natural rubbers,
they are high enough for the more crystalline polymer to
have useful properties.[10] Gutta could have large-scale
applications as both a thermoplastic and a thermosetting
resin if it were available at prices competitive with NR
and synthetic polymers.

In analytical screening of plants, infrared spectroscopy (IR) is relied on for routine detection of isoprene

TABLE 3

Weight Average Molecular Weight ($\overline{M}_w$ and Molecular Weight Distribution (MWD) for Natural Rubbers

Genus-species	Family[a]	Common Name	M_w $\overline{X}$ 10^{-3}	MWD $(\overline{M}_w/\overline{M}_n)$
Hevea brasiliensis Mull. arg.	EUP	Rubber tree	1310	5.2
Parthenium argentatum A. Gray	COM	Guayule	1280	6.1
Pycnanthemum incanum (L.) Michx.[b]	LAB	Mountain mint	495	4.0
Lamiastrum galeobdolon (L.) Ehrend. and Polatsch.	LAB	Yellow archangel	423	4.5
Monarda fistulosa L.	LAB	Wild bergamont	419	3.1
Vernonia fasciculata Michx.	COM	Iron weed	417	3.7
Symphoricarpos orbiculatus Moench	CAP	Coral berry	367	6.1
Sonchus arvensis L.	COM	Sow thistle	333	2.7
Xylococcus bicolor Nutt.	ERI	Two-color woodberry	333	2.8
Melissa officinalis L.	LAB	Balm	316	4.2
Lonicera tatarica L.	CAP	Tartarian honeysuckle	248	3.8
Silphium integrifolium Michx.	COM	Rosinweed	283	3.1
Helianthus hirsutus Raf.	COM	Hirsute sunflower	279	3.1
Cirsium vulgare (Savy) Ten.	COM	Bull thistle	266	3.1
Cacalia atriplicifolia L.[b]	COM	Pale Indian plantain	265	4.0
Euphorbia glyptosperma Engelm.	EUP	Ridgeseed Euphorbia	264	2.7
Monarda didyma L.	LAB	Oswego tea	263	3.2
Triosteum perfoliatum L.	CAP	Tinker's weed	240	3.8
Solidago altissima L.	COM	Tall goldenrod	239	3.0
Cirsium discolor (Muhl.) Spreng.	COM	Field thistle	238	3.1
Solidago graminifolia (L.) Salisb.	COM	Grass-leafed goldenrod	231	3.4
Apocynum cannabinum L.	APO	Indian hemp	216	2.7
Polymnia canadensis L.	COM	Leafy cup	206	3.5
Gnalphalium obtusifolium L.	COM	Fragrant cudweed	206	2.8

continued--

TABLE 3 -- continued

Genus-species	Family[a]	Common Name	$M_w \times 10^{-3}$	MWD ($\overline{M}_w/\overline{M}_n$)
Silphium terebinthinaceum Jacq.	COM	Prairie dock	197	3.6
Euphorbia pulcherrima	EUP	Poinsettia	197	2.7
Asclepias incarnata L.	ASC	Swamp milkweed	185	2.4
Grindelia squarrosa (Pursh.) Duval	COM	Tarweed	173	2.8
Vernonia altissima Nutt.	COM	Ironweed	167	2.7
Solidago rigida L.[b]	COM	Stiff goldenrod	164	3.1
Euphorbia corollata L.	EUP	Flowering spurge	163	3.3
Helianthus grosseserratus Martens	COM	Sawtooth sunflower	160	3.8
Elaegnus multiflora Thunb.	ELA	Cherry Elaegnus	156	2.9
Rudbeckia laciniata L.	COM	Sweet coneflower	151	2.2
Pycnanthemum virginianum (L.) Durand & Jackson	LAB	Mountain mint	147	2.8
Campsis radicans (L.) Seem. exBur.	BIG	Trumpet creeper	146	4.4
Chenopodium album L.	CHE	Lambsquarter	145	3.5
Monarda punctata L.	LAB	Horsemint	143	4.0
Apocynum androsaemifolium L.	APO	Spreading dogbane	140	2.9
Asclepias tuberosa L.	ASC	Butterfly weed	134	2.5
Nepeta cataria L.	LAB	Catnip	132	2.9
Teucrium canadense L.[b]	LAB	American germander	130	3.8
Solidago ohioensis Riddell	COM	Ohio goldenrod	127	2.4
Artemisia vulgaris L.	COM	Common mugwort	127	2.6
Aster laevis L.	COM	Smooth aster	125	2.3
Asclepias syriaca L.	ASC	Common milkweed	120	3.1
Artemisia abrotanum L.	COM	Southernwood	120	2.4
Campanula americana L.[b]	CAM	Tall bellflower	113	2.4
Centaurea vochinensis Bernh.	COM	Knapweed	111	2.4
Physotegia virginiana (L.) Benth.	LAB	Obedient plant	109	3.1

continued--

TABLE 3 -- continued

Genus-species	Family[a]	Common Name	M_w X 10^{-3}	MWD $(\overline{M}_w/\overline{M}_n)$
Verbena urticifolia L.	VER	White vervain	109	2.2
Euphorbia cyparissias L.	EUP	Cypress spurge	107	1.8
Ocimum basilicum L.	LAB	Purple basil	107	3.3
Asclepias hirtella (Pennell) Woodson	ASC	Milkweed	102	2.2
Achillea millefolium L.	COM	Yarrow	98	2.3
Phyla lanceolata (Michx) Greene	VER	Frog fruit	97.2	2.2
Gaura biennis L.	ONA	Gaura	93	2.4

[a] Code, family: APO, Apocynaceae; ASC, Asclepiadaceae; BIG, Bignoniacea; CAM, Campanulaceae; CAP, Caprifoliaceae; CHE, Chenopodiaceae; COM, Compositae, ELA, Elaegnaceae; ERI, Ericaceae; EUP, Euphorbiaceae; GRA, Gramineae; LAB, Labiateae; ONA, Onagraceae; VER, Verbenaceae.

[b] Previously selected as a promising source of natural rubber on the basis of botanical characteristic and yield of rubber and hydrocarbons.

TABLE 4

Grass Gutta Polymers

Plant source	Month/day collected	Content of hydrocarbon polymer, %	Weight-average molecular weight $\overline{M}_w \times 10^{-3}$	Molecular weight distribution, $(\overline{M}_w/\overline{M}_n)$	C^{13}_{NMR}, characterization
Agropyron repens	8/11/77	1.72	111	2.4	trans-1,4-polyisoprene
Elymus canadensis	7/28/77	1.28	116	2.4	trans-1,4-polyisoprene
Elymus canadensis	9/28/77	1.48	176	2.8	trans-1,4-polyisoprene
Elymus canadensis[a]	10/4/78	1.86	144	2.6	---
Leersia oryzoides	6/29/77	0.63	---	---	---
Leersia virginica	9/1/77	0.67	123	2.7	trans-1,4-polyisoprene
Phalaris canariensis	9/1/77	1.15	76.8	2.7	trans-1,4-polyisoprene
Phalaris canariensis	9/28/77	1.22	123	3.5	trans-1,4-polyisoprene

[a] Average values from 30 cultivated plots harvested before maturity (seeded late). Standard deviations in these measurements were respectively 0.40% gutta content, 16 $\overline{M}_w$, 0.4 $\overline{M}_w/\overline{M}_n$.

polymers and gutta is distinguished from rubber chiefly in
that it readily crystallizes upon standing at room
temperature[2,3,6] (Figure 7). However, this procedure would
not readily distinguish between mixed cis-trans-1,4-
polyisoprenes and the pure cis polymers. When the
polyisoprene content is high enough to merit further
characterization, nuclear magnetic resonance (NMR) is used
to determine detailed chemical structure; but so far mixed
isomers have not been identified.

The older traditional methods of detecting rubber in
plant extractives could scarcely distinguish between rubber
and gutta.[16] Thus, it is possible that there are species
producing mixed cis-trans-polyisoprenes. However, such
details of structure are immaterial from a practical view-
point for low-molecular weight polyisoprenes considered
only as a hydrocarbon source.

LIGNOCELLULOSE CONVERSION PRODUCTS

Since considerable research is in progress and several
recent symposiums and reviews have been given on biomass
conversion to fuels and feedstocks, we have limited our
discussion of this subject to the extractive-free residues
from botanochemical processing of potential new crops.
These residues would be the main product by weight of a
botanochemical extraction process and would be available as
"flakes" for subsequent chemical or microbiological conversion to a wide variety of secondary botanochemical (Figure
8).

The minimum value for extractive-free residue would be
about $50/ton (July 1979) as a solid fuel. There are
several thermal-pyrolytic-catalytic routes to gaseous and
liquid fuels or petroleum substitutes from
lignocellulose.[1,11,13,14,18] Anaerobic fermentation is
another alternative route to gaseous fuel, i.e., methane.

Extractive-free residues would be especially attractive
sources of lignocellulose for saccharification to sugars
for fermentation to fuel alcohol, as suggested also in
Figure 1. Pentosans in the residue can either be converted
to 5-carbon sugars for fermentation or to furfural.

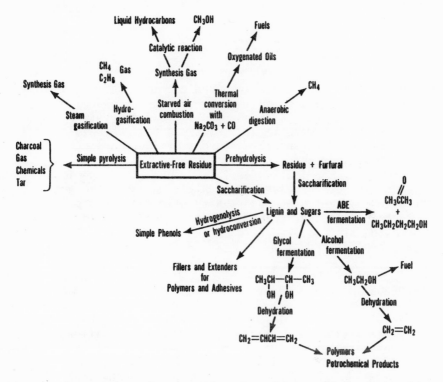

Fig. 8   Conversion of extractive-free lignocellulosic residues
         to botanochemicals.

Other secondary botanochemicals that could be produced economically by fermentation of low-cost sugars include acetone, butanol, and 2,3-butanediol. Alternately, many of these residues could be used for nonbotanochemical applications. The process for flaking whole-plant produce can be conducted with minimum damage to the fiber so that the residue can be used for paper- or board-making. Also, any of several treatments for increasing the digestibility and the protein content of the residues would yield a valuable cattle feed.

CONCLUSION

Green plants are solar-powered chemical factories which convert CO_2 and H_2O to an immense variety of energy-rich organic compounds. As green plants grow, they efficiently store energy and materials. Crops can be developed to help meet the need for renewable sources of fuels and feedstocks, and conceptually botanochemical production systems are economically feasible. Thus, phytochemical research is now perceived as highly relevant, even urgent. Moreover, a developing opportunity exists for phytochemists to contribute to the creation of new crops, new industrial processing systems, and a new green-crop-based synthetic botanochemical industry.

REFERENCES

1. Antal, M.J., Jr. 1979. The effects of residence time, temperature and pressure on the steam gasification of biomass. Symposium, Biomass as a Nonfossil Fuel Source. 1979 Pacific Chemical Conference, Joint Meeting American Chemical Society and Chemical Society Japan, Honolulu, Hawaii, April 1-6.

2. Buchanan, R.A., I.M. Cull, F.H. Otey, and C.R. Russell. 1978a. Econ. Bot. 32:131,146.

3. Buchanan, R.A., F.H. Otey, C.R. Russell and I.M. Cull. 1978b. J. Am. Oil Chem. Soc. 55:657.

4. Buchanan, R.A. and F.H. Otey. 1979a. Multi-use oil- and
    hydrocarbon-producing crops in adaptive systems for
    food, material and energy production. In: Larrea: A
    Vast Resource of the American Deserts (E. Campos-
    Lopez and T.J. Mabry, eds.). Consejo Nacional de
    Ciencia y Technologia, D.F. Mexico.

5. Buchanan, R.A. and F.H. Otey. 1979b. Multi-use botano-
    chemical crops. Economic assessment and feasibility
    study. Symposium, Fuels and Chemical Feedstocks from
    Renewable Resources. 11th Central Regional Meeting,
    American Chemical Society, Columbus, Ohio, May 7-9.

6. Buchanan, R.A., C.L. Swanson, D. Weisleder, and I.M.
    Cull. 1979c. Phytochemistry 18:1069.

7. Gushee, D.E. 1978. Ind. Eng. Chem., Prod. Res. Dev.
    17:93.

8. Haag, W. 1979. Mobile Research and Development
    Corporation, Princeton, New Jersey, private
    communication, July.

9. Hemmingway, R.W. 1977. Bark, a renewable resource of
    specialty chemicals. Symposium, Chemicals from
    Renewable Resources. W.R. Grace and Company,
    Columbia, Maryland, November 14.

10. Kent, E.G. and F.B. Swinney. 1966. Ind. Eng. Chem.,
    Prod. Res. Dev. 5:134.

11. Kuester, J.L. 1978. Liquid fuels from biomass. AIAA/
    ASERC Conference on Solar Energy: Technology Status.
    Phoenix, Arizona, November 27-29.

12. Lipinsky, E.S. 1978. Science 199:644.

13. Molton, P.M., T.F. Demmitt, J.M. Donovan and R.K.
    Miller. 1978. Mechanism of conversion of cellulosic
    wastes to liquid fuels in alkaline solution.
    Symposium, Energy from Biomass and Wastes. Institute
    of Gas Technology, Chicago, Illinois, August 14-18.

14. Peart, R.M. 1978. Process gas from organic wastes: Corn
    cobs. Symposium, I Seminaro Sobre Energia de
    Biomasssas no Nordeste. Universidade Federal do
    Ceara, Fortalenza, Brazil, August 15-18.

15. Pizzi, A. and D.G. Roux. 1978. J. Appl. Polym. Sci. 22:
    1945.

16. Polhamus, L.G. 1967. ARS 34-74 Bulletin, Agricultural
    Research Service, USDA.

17. Swanson, C.L., R.A. Buchanan, and F.H. Otey. 1979. J.
    Appl. Polym. Sci. 23:743.

18. Waterman, W.W. and D.L. Klaas. 1976. Biomass as a long
    range source of hydrocarbons. Symposium, Shaping the
    Future of the Rubber Industry. 110th Meeting Division
    of Rubber Chemistry, ACS, San Francisco, California,
    October 5-8.

19. Weisz, P.B., W.O. Haag, and P.G. Rodewald. 1979.
    Science 206:57.

20. 95th U.S. Congress. 1978. Public Law 95-592. Native
    Latex Commercialization and Economic Development Act.

Chapter Two

ANTITUMOR AGENTS FROM HIGHER PLANTS*

RICHARD G. POWELL AND CECIL R. SMITH, JR.

Northern Regional Research Center, Agricultural
Research, Science and Education Administration,
U.S. Department of Agriculture
Peoria, Illinois 61604

INTRODUCTION

Various plant preparations have been recommended for
treatment of warts, tumors and cancerous growths throughout
recorded history. In a classic survey of the available
literature, Hartwell[1] identified over 3,000 plants reported
to be of use in the treatment of these maladies. Numerous
plant extracts have been subjected to detailed testing and
fractionation studies in recent years and a wide variety of
new active compounds have been isolated. Many of these new
materials show promising antitumor activity in animal test
systems and in preliminary clinical trials.

Prior to 1960, only a few well-characterized compounds
from higher plants were known to possess antitumor
activity. These earlier active compounds included
podophyllotoxin and related lignans from Podophyllum
peltatum L. (May apple or mandrake); demecolcine,

* The mention of firm names or trade products does not
imply that they are endorsed or recommended by the U.S.
Department of Agriculture over other firms or similar products not mentioned.

23

N-deacetyl-N-methylcolchicine, from Colchicum autumale L.
(autumn crocus); and the alkaloids vinblastine and
vincristine from Vinca rosea L. (Catharanthus roseus G.
Don). The two Vinca alkaloids are now used routinely in
clinical treatment of cancer and vincristine is considered
to be the drug of choice in treating acute leukemias in
children. This alkaloid is a component of virtually every
combination chemotherapy regimen for cancer used today.
Vinblastine is widely used for the management of
methotrexate-resistant Hodgkins' disease and for
choriocarcinoma.[2] These examples clearly demonstrate that
antineoplastic substances useful in effective management of
human disease occur in plants.

Podophyllotoxin   Demecolcine

Vinblastine
R = CH₃

Vincristine
R = CH=O

    An intensive search for other plant products having
antitumor activity began in 1957, shortly after establish-
ment of the Cancer Chemotherapy National Service Center
(CCNSC) within the National Cancer Institute (NCI). A
brief history of the program, currently under direction of
the Natural Products Branch, Developmental Therapeutics

Program, Division of Cancer Treatment, NCI, has been presented by Shephartz.[3] Although many other sources of plant
material have been utilized, the bulk of plant collecting
has been carried out through a cooperative agreement with
the United States Department of Agriculture (USDA).[4]
Bioassay methods used in the testing of plant extracts for
antitumor activity have been summarized by Abbott,[5] and
Geran et al.[6] have published detailed protocols for the
screening of natural products against animal tumors.
Primary screening of extracts has varied over the years but
typically involves an in vitro test for cytotoxicity in the
KB cell culture and an in vivo test against P-388 lympho-
cytic leukemia (PS) or L-1210 leukemia (LE) in mice. Other
test systems of particular importance include Walker
carcinosarcoma-256 (WM), Lewis lung carcinoma (LL), colon
adenocarcinoma 38 (C8), and B-16 melanotic melanoma (B1).
Purified active compounds are obtained by systematic
fractionation of extracts that show promise in the preliminary screen. Many of the newly isolated compounds are
unique, and often the discovery of a new structural type
leads to isolation or synthesis of an entire series of
related compounds. Progress in the search for antitumor
compounds from plants has been reviewed periodically over
the years, and several excellent summaries are
available.[7-12]

This report will emphasize antitumor active compounds
that have undergone clinical trials or are being considered
for such investigation, as well as some newer active compounds of special interest; also included will be some
observations on the screening of seed extracts from our own
laboratory. Widely studied classes of compounds of limited
clinical interest, e.g., sesquiterpene lactones, cucurbitacins, saponins, and cardenolides, will be mentioned only
where appropriate.

TUMOR INHIBITORS OF EARLY INTEREST

By 1969, several hundred compounds had been discovered
that showed activity in experimental tumor systems. Of the
wide variety of structural types isolated from plants, the
alkaloids are undoubtedly the most interesting.

Thalicarpine

Emetine

Tetrandrine

Two dimeric benzylisoquinoline alkaloids, thalicarpine and tetrandrine, are of substantial interest as antitumor agents. Their structural elucidation is due primarily to the efforts of Kupchan and co-workers. Thalicarpine,[13] isolated from Thalictrum dasycarpum Fisch. (Ranunculaceae) and tetrandrine[14] from Cyclea peltata Diels (Menisperma-ceae), both exhibit marked activity against Walker carcinosarcoma-256 (WM) in rats and were selected for preclinical evaluation on this basis. A synthesis of thalicarpine has been described[15] and the structure of tetrandrine was determined by x-ray analysis.[16] Clinical experience with thalicarpine has been disappointing since central nervous system (CNS) and cardiovascular toxicities were encountered in man at doses below those required for antitumor activity.[17] Preclinical toxicology demonstrated that tetrandine also has a number of serious side effects.[18]

Emetine, an isoquinoline alkaloid isolated from Cephaelis acuminata (Rubiaceae), exhibits activity in both the P-388 and L-1210 leukemia systems. Emetine is an older drug, widely used as an amoebicide. Clinical trials demon-strated some activity against lung carcinoma at high dose levels, but no beneficial results were observed against several other malignancies.[19-20]

Investigation of a Chinese tree, Camptotheca acuminata Decne. (Nyssaceae) by Wall and coworkers revealed another

important series of alkaloids. These include camptothecin[21] and its 10-hydroxy and 10-methoxy derivatives.[22]
Camptothecin is exceptionally active in the L-1210 and
P-388 leukemia systems and shows considerable activity
against a variety of other animal tumors, including Walker
256 carcinoma, Lewis lung carcinoma, and B-16 melanotic
melanoma. The insolubility of camptothecin was a problem
in preclinical evaluation. Fortunately, the lactone ring-
opened sodium salt (camptothecin sodium) is also active,
both orally and intravenously, and most of the animal and
clinical studies have been carried out on this water-
soluble derivative. Early enthusiasm for camptothecin has
been dampened by the frequency and severity of its adverse
side effects and apparent inactivity in clinical
trials.[23-25] Larger amounts of camptothecin are available
from two other sources, Mappia foetida (Olinaceae)[26] and
Ophiorrhiza mungos L. (Rubiaceae).[27] Numerous synthetic
routes have appeared for the preparation of camptothecin
and related alkaloids.[28-30]

Camptothecin   R = H
10-Hydroxycamptothecin R = OH
10-Methoxycamptothecin R = OCH₃

Camptothecin Sodium

Ellipticine, 9-methoxyellipticine, and olivacine,
obtained from several Ochrosia and Aspidosperma species
(Apocynaceae), are of considerable interest due to their
activity against L-1210 and a variety of other tumors. The
more important of these is probably 9-methoxyellipticine,
first isolated from Ochrosia maculata.[31] French workers
have studied a number of ellipticine derivatives and find
that 9-hydroxyellipticine and 9-aminoellipticine[32] retain
activity. Ellipticine exhibits troublesome side effects,
including hemolysis and cardiac depression in monkeys.[33]
Considerable synthetic work remains to be done in this
important series of pyridocarbazole alkaloids.

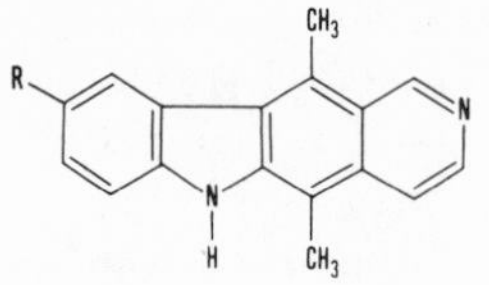

Ellipticine   R = H
9-Methoxyellipticine   R = OCH₃
9-Hydroxyellipticine   R = OH
9-Aminoellipticine   R = NH₂

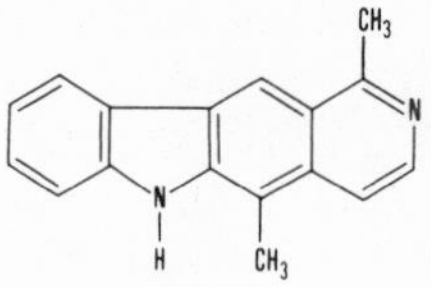

Olivacine

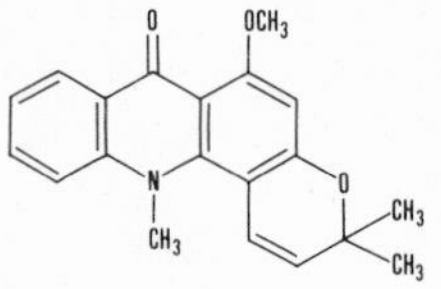

Acronycine

Svoboda and coworkers[34] isolated an acridone alkaloid,
acronycine, from the bark of an Australian tree, Acronychia
baueri Schott (Rutaceae). Acronycine has an unusually
broad spectrum of antitumor activity and is effective when
administered orally or subcutaneously. As with camptothecin, water insolubility has been a problem in formula-
ting this drug. Structure-activity relationships of
acronycine analogs have been investigated[35-36] and several
syntheses of acronycine are available.[37] Of the analogs
studied, only acronycine shows appreciable activity.

Phenanthroindolizidine alkaloids with cytotoxic or
L-1210 activity occur in Tylophora and other genera of the
Asclepiadaceae.[38] The more important of these include
tylophorine, tylophorinine and tylocrebrine. Clinical
trials of tylocrebrine, from T. crebriflora S. T. Blake[39]
were not promising due to unmanageable CNS effects. This
group of alkaloids is exceptional in that activity is
retained in many of the analogs; a more usual situation is
that slight changes in structure eliminate antitumor
activity.

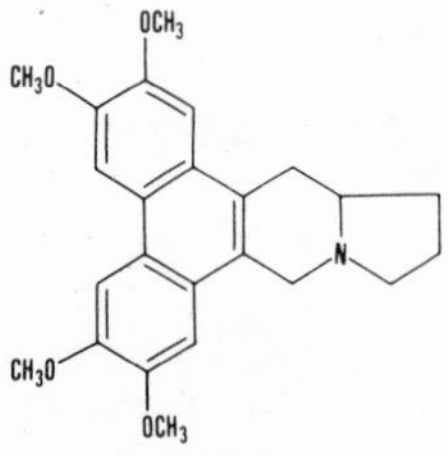

Tylophorine

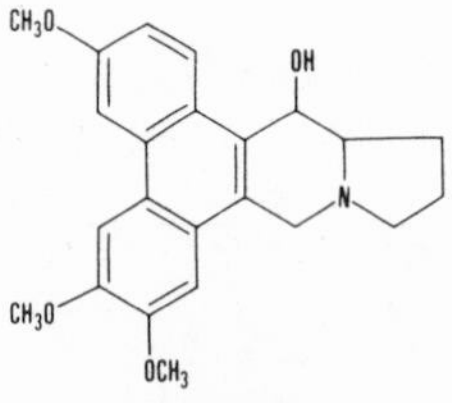

Tylophorinine

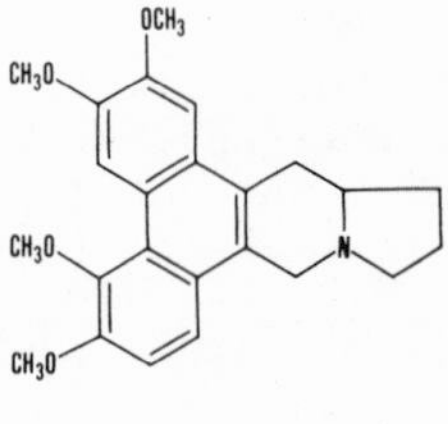

Tylocrebrine

Monocrotaline   Indicine N-oxide

Pyrrolizidine alkaloids occur most frequently in the families Compositae, Boraginaceae and Leguminosae. As a group, they are considered teratogenic and strongly hepatotoxic; furthermore, they are high on the list of suspected or proven carcinogens.[40] Paradoxically, several of this group, including monocrotaline[41] and indicine N-oxide,[42] are noted for their significant antitumor activity. Indicine N-oxide is active against several animal tumors, and hepatotoxicity does not appear to be a serious clinical problem with this alkaloid.[9]

Also prominent among the early antitumor agents from plants were a number of sesquiterpene lactones, a few quinones, and a variety of saponins, cardiac glycosides, cucurbitacins, expoxides, and polypeptides.[7]

Elephantopin   Vernolepin

Sesquiterpene lactones, such as elephantopin[43] and vernolepin,[44] are common constituents of most genera of the Compositae and are occasionally found in other families. Some fifty of these were discussed as tumor inhibitors in the 1969 review by Hartwell and Abbott[7] and a wealth of literature is available on the isolation, biological activity and synthesis of these compounds. Kupchan et al. elaborated on their structure-cytoxicity relationships,[45] and Rodriguez and coworkers have summarized the biological activities of this complex group.[46] The cytotoxic sesquiterpenes normally contain an exocyclic α,β-unsaturated lactone, and it appears that the same structural features

responsible for antitumor activity are responsible for
allergic contact dermatitis in man.[47] Thus, the sesquiterpene lactones have not been attractive candidates for
clinical trials, and their potential as antitumor agents
appears to be severely limited.

Lapachol

Lapachol, well known as a constituent of the wood of
several species of the Bignoniaceae, was isolated as the WM
active component of Stereospermum suaveolens (Roxb.) DC.
A tea prepared from members of this family has enjoyed some
popularity in Brazil as a cancer treatment, and lapachol is
more active when administered orally than intraperitone-
ally. Lack of toxicity has permitted large oral doses in
clinical trials; however, some anticoagulant activity was
noted and blood levels high enough to show therapeutic
effects were not realized.[48-49].

NEWER TUMOR INHIBITORS

Screening of plant materials has intensified during the
past decade, with a corresponding increase in the number of
potentially useful antitumor agents available. Attention
has focused on benzophenanthridine and Cephalotaxus alkaloids, a variety of complex diterpenes, ansa macrolides,
and a few other miscellaneous compounds.

Fagaronine

Nitidine chloride

Two benzophenanthridine alkaloids, fagaronine and nitidine (as their chlorides), show excellent activity in the

P-388 leukemia system, and are the subjects of intensive investigations. These and related alkaloids occur in Fagara and Zanthoxylum (Rutaceae) species.[50-51] Two syntheses of fagaronine[52-53] have been reported, and the structure-activity relationships of many related benzo-phenanthridine derivatives were discussed by Wall and Wani[10] and by Stermitz et al.[54]

R   OH O
(CH₃)₂C(CH₂)ₙ C—C   OCH₃
              CHR'
              CO₂CH₃

HO

OCH₃

Harringtonine
R = OH, R' = H, n = 2
Isoharringtonine
R = H, R' = OH, n = 2
Homoharringtonine
R = OH, R' = H, n = 3
Deoxyharringtonine
R = H, R' = H, n = 2

Cephalotaxine

Cephalotaxus (Cephalotaxaceae), a small genus of yew-like evergreen trees and shrubs native to southeastern Asia, is receiving world-wide attention as a source of promising antitumor agents. Studies in our laboratory demonstrated that four alkaloids, harringtonine, iso-harringtonine, homoharringtonine, and deoxyharringtonine, were responsible for the exceptional P-388 and L-1210 activity of Cephalotaxus harringtonia extracts.[55] Homoharringtonine is also active in the B1 melanoma and C8 colon adenocarcinoma systems.[56] Cephalotaxine, the inactive parent alkaloid, has been synthesized[57-58] and synthetic routes to deoxyharringtonine,[59] harringtonine,[60-62] and some active analogs are available.[63] Research on the Cephalotaxus alkaloids has recently been reviewed.[64] Meager supplies of the harringtonines in the United States have delayed clinical trials, although encouraging reports are available from the Peoples Republic of China.[65-66] This novel group of alkaloids is among the most promising of the newer antitumor agents available from higher plants.

Two nitrogen-containing esters of a taxane diterpene, baccatin III, are under investigation. Taxol, isolated by Wani and coworkers as a constituent of Taxus brevifolia (Taxaceae),[67] is highly active in the L-1210 and P-388 leukemia systems and is being prepared in quantity for clinical trials. A second member of this series, cephalomannine[68] was recently isolated in the authors' laboratory from plant material initially identified as <u>Cephalotaxus mannnii</u>.* Taxol and cephalomannine exhibit similar activity in animal tests; however, formulation of these could present problems because both compounds are easily hydrolyzed and are sparingly soluble in aqueous systems.

Baccatin III   R = OH

Taxol

Cephalomannine

Bruceantin

Ailanthinone

6 α-senecioyloxychaparrinone

Esters of the tigliane, daphnane, and ingenane diterpenes are noted for their skin irritant and cocarcinogenic effects.[74] These compounds occur widely in toxic species of the Euphorbiaceae and Thymelaeaceae. Ironically, many of the group, particularly orthoesters of the daphnane series, exhibit potent antileukemic activity. There is considerable theoretical interest in the structural features of this series that are conducive to antitumor activity as opposed to those that are responsible for irri-

* In view of our chemical results, the botanical classification of our plant material is being re-evaluated.

Gnididin
R = COCH=CHCH=CH(CH₂)₄CH₃

Mezerein
R = COCH=CHCH=CHC₆H₅

Gnidimacrin

tant properties. Representative of the daphnane tumor
inhibitors are gnididin from <u>Gnidia lamprantha</u> Gilg,[75],
gnidimacrin from <u>G. subcordata</u> (Meissn.) Engl.[76] and
mezerein from Daphne mezereum L.[77]

The family Simaroubaceae contains many diterpenoid
bitter principles, termed quassinoids[69] or simaroubolides,
which show interesting antitumor activity in animal tests.
Quassinoids were originally investigated for their
antiamoebic activity. Bruceantin, from <u>Brucea antidysen-</u>
<u>terica</u> Mill.,[70] is certainly one of the most important com-
pounds in this series and is currently in clinical trial.
Other representative quassinoids include ailanthinone from
<u>Ailanthus excelsa</u> Roxb.[71] and 6α-senecioyloxychaparrinone
from <u>Simaba multiflora</u> A. Juss.[72] Wall and Wani have commented on some structural features necessary for activity
in the quassinoids.[73]

Maytansine
R = R' = H

Colubrinol
R = CH₃; R' = OH

Ansa macrolides or maytansinoids are among the more
exciting groups of antitumor compounds isolated from plants
in recent years. Maytansine, from <u>Maytenus ovatus</u> Loes
(Celastraceae),[78] and colubrinol, from <u>Colubrina texensis</u>
Gray (Rhamnaceae),[79] are representative of this series.
The maytansinoids exhibit broad spectrum activity in the
µg/kg range. Early work on maytansine, the most widely
studied of this series, was hampered by the exceptionally

low yields (0.2 mg/kg) of this compound from Maytenus
species. Higher yields of maytansine (12 mg/kg) have
been obtained from Putterlickia verrucosa Szyszyl.
(Celastraceae).[80] Requirements for activity in the maytansinoids include an ester function at C-3.[81] In a clinical
trial, toxicity was manifested by profound weakness,
nausea and prolonged diarrhea to the extent that some
patients refused further treatment. Some beneficial antitumor responses were noted, however, and maytansine is
undergoing further evaluation.[82]

Sesbanine

    Three legumes toxic to livestock and usually regarded
as Sesbania spp. are found along the Eastern Coastal Plains
from Carolina to Texas.[83] Investigations in our laboratory
revealed that ethanolic extracts of Sesbania vesicaria
(Jacq.) Ell., S. punicea (Cav.) Benth., and S. drummondii
(Ryd.) Cory. seeds were cytotoxic and highly active against
P-388 leukemia.[84] Further work demonstrated that some of
the cytotoxic activity of S. drummondii was due to
sesbanine.[85] It is not yet clear if sesbanine is also PS
active as the extremely low yield obtained (≈50 mg from
450 kg of seed) has precluded extensive in vivo testing.
Sesbanine has a previously unreported and highly unusual
spirocyclic structure based on the 2,7-naphthyridine
nucleus and has captured the imagination of several groups
of synthetic chemists. Further testing of this novel antitumor agent will be carried out when suitable quantities
are available.

    Other plant-derived antineolastic agents of current
clinical interest include triptolide and tripdiolide from
Tripterygium wildfordii Hook F. (Celastraceae)[86] and a
series of macrocyclic tricothecenes, including baccharin,

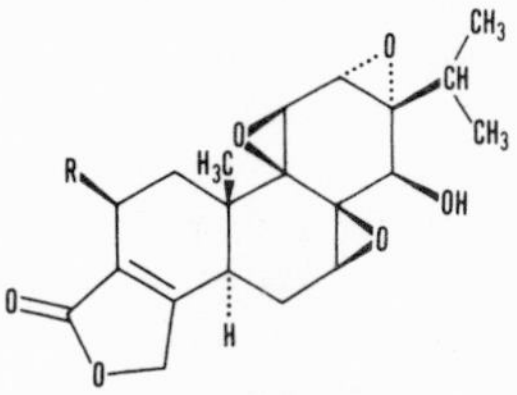

Triptolide
R = H
Tripdiolide
R = OH

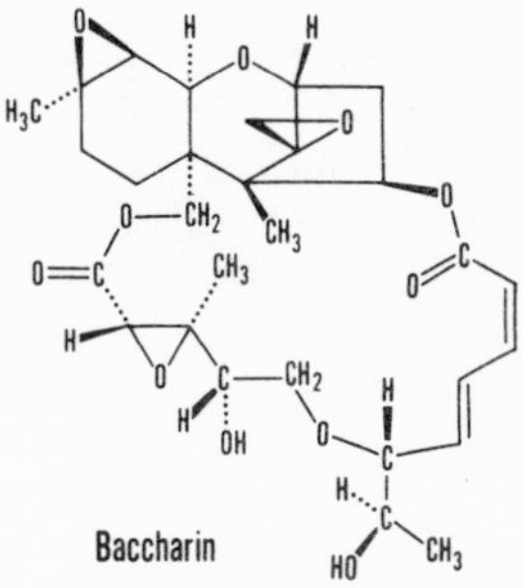

Baccharin

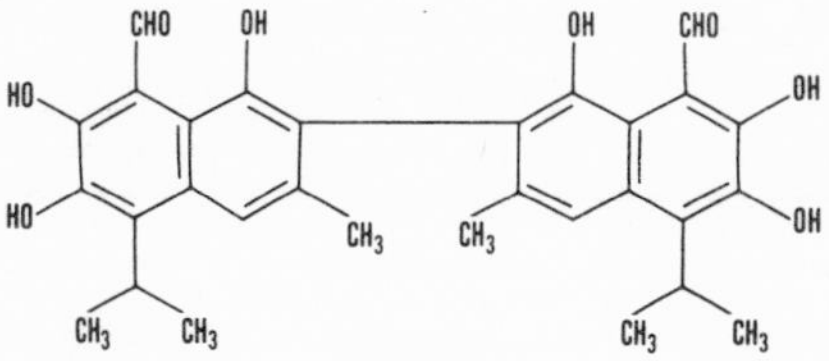

Gossypol

from Baccharis megapotamica Spreng (Asteraceae).[87] The
tricothecene mycotoxins are an important group of mold
metabolites, and work is being done with fungi found
growing on the Baccharis plant in order to determine if
baccharin is also of microbial origin. Other
tricothecenes, including anguidin and T-2 toxin, are
undergoing tests at NCI.[88] It is of interest to note that
gossypol, long known as a toxic constituent in the
Malvaceae, was isolated as the tumor inhibitory principle
of Montezuma speciosissima Sesse and Moc.[89] Gossypol is
currently a leading candidate for a new male oral contra-
ceptive agent,[90] and extensive clinical trials for this
purpose are being conducted in the Peoples Republic of
China.

SCREENING OF SEED EXTRACTS FOR ANTITUMOR ACTIVITY

Random screening of seeds from our laboratory began in
1958 when eighteen aqueous, ethanol and light petroleum
extracts were submitted to the CCNSC. Emphasis was placed
on seed extracts because very few seed extracts were among
the large number of plant extracts being screened. Seeds
often accumulate physiologically active materials, and

several thousand seed samples from many areas of the world
were readily available at the Northern Regional Research
Center. By 1963, a total of 221 samples had been screened
with heavy emphasis on freeze-dried aqueous extracts. Nine
of these original extracts, including an aqueous extract of
Cephalotaxus harringtonia, gave confirmed activity in one
or more of the tumor systems in use at that time. More
recent extracts have been prepared by Soxhlet extraction of
defatted seed meals with 95% ethanol. We have now submitted over 1,100 samples to the screen, including representa-
tives of 722 genera and 894 species. Approximately 6% of
these demonstrate some activity that might warrant further
study.

Antileukemic (PS) and cytotoxic (KB) activity of our
more active seed extracts are listed in Tables 1 and 2.
Cytotoxicity in the KB cell culture seems to be an indicator of useful antitumor activity only if the material also
confirms in an in vivo screen. Most important active compounds have been obtained from extracts yielding an initial
T/C of 150% or more in the PS leukemia system, and any
extract having at least this level of activity is a prime
candidate for further fractionation. Thus, our Maytenus,
Cephalotaxus, Trewia, Sesbania and Ailanthus extracts were
of particular interest. However, the probable presence of
maytansine in the Maytenus extract and of ailanthinone in
the Ailanthus extract eliminated these two from further
fractionation studies in our laboratory, since extracts
from these genera already were being studied by other
investigators.

Of all the active extracts covered by the data in Table
1, the most exciting results have been realized from our
studies of Cephalotaxus harringtonia. The Cephalotaxus
alkaloids, harringtonine and related cephalotaxine esters,
offer a broad spectrum of activity, a high therapeutic
index, are reasonably stable, and possess favorable solubi-
lity properties. These alkaloids are present in all parts
of the plant and, with considerable variation in quantity,
they are found in all Cephalotaxus species and varieties we
have examined[91] except C. mannii.[68]   The Chinese appear to
favor C. hainanensis as a source of these materials.

TABLE 1
Activity of Some Seed Extracts Against
PS Leukemia in Mice[a]

Species	Dose, mg/kg at maximum T/C	T/C (%)
Celastraceae		
Maytenus sp.	100	213
Cephalotaxaceae		
Cephalotaxus harringtonia[b]	10	285
Compositae		
Carelia cistifolia	400	133
Senecio abrotanifolius	400	145
Senecio orientalis	400	140
Datiscaceae		
Datisca cannabina	6	137
Euphorbiaceae		
Baliospermum montanum	200	144
Euphorbia helioscopia	100	128
Euphorbia lagascae	200	137
Trewia nudiflora	100	163
Iridaceae		
Tigridia pavonia	100	137
Leguminosae		
Sesbania drummondii	30	176
Sesbania punicea	20	205
Sesbania vesicaria	40	190
Tephrosia nyikensis	50	144
Resedaceae		
Reseda phyteumia	400	146
Scrophulariaceae		
Bellardia trixago	400	135
Simaroubaceae		
Ailanthus excelsa	400	177

[a] 95% Ethanol extracts of defatted seed unless indicated
otherwise. Materials are considered active if the survival time of animals treated (T) with them was >125% of
that of the controls (C), i.e., T/C >125%.

[b] Alkaloid concentrate

TABLE 2

Cytotoxicity of Selected Seed Extracts

in the KB Cell Culture[a]

Species	ED_{50} μg/ml	Species	ED_{50} μg/ml
Alismaceae		Lythraceae	
Alisma plantago	1.5×10^1	Heimia myrtifolia	4.0×10^0
Annonaceae		Myrtaceae	
Annona glauca	3.6×10^0	Eucalyptus sieberiana	7.6×10^0
Apocynaceae		Onagraceae	
Carissa grandiflora	2.2×10^0	Epilobium hirsutum	8.0×10^0
Nerium indicum	3.2×10^0	Jussiaea leptocarpa	1.6×10^1
Strophanthus kombe	2.7×10^{-2}	Pinaceae	
Thevetia thevetioides	5.3×10^0	Chamaecyparis lawsoniana	2.9×10^0
Celastraceae		Polygonaceae	
Maytenus sp.	4.5×10^{-1}	Ruprechtia salicifolia	2.2×10^1
Cephalotaxaceae		Ranunculaceae	
Cephalotaxus harringtonia	3.9×10^0	Helleboris vesicarius	4.5×10^{-1}
Cucurbitaceae		Rosaceae	
Maximowiczia sonorae[b]	2.4×10^0	Fallugia paradoxa	6.6×10^0
Datiscaceae		Geum urbanum	1.1×10^1
Datisca cannabina	2.7×10^0	Rutaceae	
Euphorbiaceae		Ruta bracteosa	6.0×10^1
Trewia nudiflora	3.1×10^{-2}	Scrophulariaceae	
Guttiferae		Digitalis ferruginea	6.5×10^0
Mammea americana[b]	2.3×10^0	Digitalis purpurea	8.3×10^0
Hypoxidaceae		Digitalis thaspi	7.4×10^0
Hypoxis aurea	3.0×10^0	Thymelaeaceae	
Leguminosae		Daphne pontica	2.0×10^1
Arthrosamanea polyantha	1.8×10^1		
Erythrophleum guineense	4.6×10^0		
Goldmannia foetida	1.0×10^1		
Sesbania drummondii	3.2×10^1		
Sesbania punicea	6.0×10^0		
Sesbania vesicaria	1.1×10^1		

[a] 95% Ethanol extracts of defatted seeds unless indicated otherwise.

[b] Pentane-hexane extract.

Three Sesbania spp., S. drummondii, S. punicea, and S. vesicaria, appear next in importance and, as discussed earlier, we obtained an active constituent, sesbanine, from S. drummondii.[85] The presence of sesbanine in S. punicea and S. vesicaria has yet to be demonstrated, although crude extracts of all three species exhibit similar activities. The Trewia nudiflora extract is in an advanced stage of fractionation.

Datiscoside   Montanin

A review of phytochemical precedents allows speculation concerning the nature of active components likely to be present in other extracts listed in Table 1. Activity of the two Senecio species, S. abrotanifolius and S. orientalis, is probably due to pyrrolizidine alkaloids, while Datisca cannabina probably contains datiscoside, known as a component of D. glomerata Baill,[92] or some related cucurbitacin. Montanin was recently isolated from Baliospermum montanum,[93] and similar diterpenes may be expected in other members of the Euphorbiaceae. The KB cytotoxicity of extracts from plants in the Apocynaceae and Scrophulariaceae, Table 2, is likely due to cardiac glycosides which occur widely in these families. Highly toxic alkaloids such as 3β-acetoxynorerythrosuamine,[94] occur in Erythrophleum spp., and the cytotoxicity of Annona glauca could be accounted for by the presence of liriodenine,[95] or similar alkaloids. However, there are limitations to this type of speculation and, unless active compounds present in each new species are fully characterized, many potentially useful drugs could be overlooked.

3β-Acetoxynorerythrosuamine   Liriodenine

TABLE 3

Comparative Activity of Selected Plant Materials

Compound tested or source of ethanol extract	% inhibition of crown-gall tumor initiation[a]	Cytoxicity KB cell culture ED_{50} $\mu g/ml$	P388 Leukemia % increase in life span (mg/kg)
Conidendrin	0/0	1.0×10^2	Inactive
Raffinose	0/0	1.0×10^2	Inactive
Akebia quinata	0/0	4.4×10^1	Inactive
Cephalotaxine	20/0	1.0×10^2	Inactive
Homoharringtonine	83/58	1.0×10^{-2b}	238 (1.0)
Isoharringtonine	60/82	1.7×10^{-1b}	172 (7.5)
Sesbania punicea	79/52	2.4×10^{0b}	105 (20)
Trewia nudiflora	29/51	3.1×10^{-2b}	63 (100)
Geum urbanum	0/0	1.1×10^{1b}	Inactive
Thevetia thevetioides	0/0	4.0×10^{-1b}	Inactive
Neriifolin	0/0	2.2×10^{-2b}	Inactive

a Duplicate trials. Test materials were dissolved in water, 0.2 mg/ml, and filtered through 0.22 μ Millipore Filters into tubes containing an equal volume of A. tumefaciens strain B6 (a 48-hr culture containing 5 X 10^9 cells/ml). 0.5 ml of these solutions were used to inoculate the potato discs.

b Active materials.

Promising results have recently been obtained using the crown-gall tumor disc bioassay as a tool in screening seed extracts for antitumor activity. This work was initiated and is being pursued by Alan Galsky of Bradley University.[96] Crown-gall is a neoplastic disease of plants which occurs in more than 60 families of dicotyledons and in many gymnosperms. The causative agents of this disease are specific strains of the gram negative bacterium Agrobacterium tumefaciens.[97] Tumors are induced by inoculating potato discs with the bacterium and, after 12 days, the tumors are counted by eye or with the aid of a low-power microscope. Test materials may be added to the potato discs simultaneously with the bacteria or up to 30 min following inoculation.

Results of our initial studies on the crown-gall system were most encouraging, as presented in Table 3. Materials known to be highly PS active in vivo gave marked inhibition of the crown-gall tumors, while those which were only KB active (cytotoxic) were completely inactive in this screen. Active extracts did not affect bacterial viability, and the amount of inhibition obtained was independent of the time of addition of test materials. In some instances, actual regression of developed tumors was observed, similar to the earlier observations of Richardson and Morré on bean leaves.[98] The crown-gall tumor disc bioassay appears to offer great potential as a rapid, inexpensive, and highly selective screening tool in the detection of antitumor activity, and further tests are underway to assess its predictive ability. It certainly could be used to eliminate many unpromising KB actives and for rapid in-house assay of fractions.

SUMMARY

Higher plants have yielded a large number of antitumor active agents over the last three decades. Often the plants yielding these compounds have been used by practitioners of traditional medicine, or as folk remedies, for conditions alleged to be cancer. The Vinca alkaloids, vinblastine and vincristine, are outstanding examples of cancer chemotherapeutic agents derived from plants. Some of the more promising active compounds, such as thalicarpine, camptothecin, acronycine, emetine, maytansine, tylocrebrine, and lapachol, have been included in clinical

trials with varying degrees of success. Newer plant products currently under evaluation in the United States and
elsewhere include baccharin, tripdiolide, homoharringtonine, taxol, bruceantin, 9-hydroxyellipticine, and indicine
N-oxide. There is every reason to believe that a variety
of clinically useful antitumor drugs may be among these
plant products or among those which remain to be
identified. Our laboratory has achieved some success in
contributing to the list of potential new drugs by
screening and fractionating seed extracts.

ACKNOWLEDGMENT

    Antileukemic and cytotoxic results were obtained through
cooperation with the National Cancer Institute and data on
the crown-gall bioassay through Dr. Alan Galsky.

REFERENCES

1. Hartwell, J.L. 1967. Plants used against cancer. A survey. Lloydia 30:379 (with ten additional installments
   ending with Lloydia 34:386).

2. Carter, S.K. and R.B. Livingston. 1976. Plant products
   in cancer chemotherapy. Cancer Treat. Rep. 60:1141.

3. Schepartz, S.A. 1976. History of the National Cancer
   Institute and the Plant Screening Program. Cancer
   Treat. Rep. 60:975.

4. Perdue, R.E., Jr. 1976. Procurement of plant materials
   for antitumor screening. Cancer Treat. Rep. 60:987.

5. Abbott, B.J. 1976. Bioassay of plant extracts for anticancer activity. Cancer Treat. Rep. 60:1007.

6. Geran, R.I., N.H. Greenberg, M.M. MacDonald, A.M.
   Schumacher and B.J. Abbott. 1972. Protocols for
   screening chemical agents and natural products
   against animal tumors and other biological systems
   (Third Edition). Cancer Chemother. Rep. Part 3, 3:1.

7. Hartwell, J.L. and B.J. Abbott. 1969. Antineoplastic
   principles in plants: recent developments in the
   field. Adv. Pharmacol. Chemother. 7:117.

8. Kupchan, S.M. 1975. Advances in the chemistry of tumor-inhibitory natural products. Rec. Adv. Phytochem. 9:167.

9. Hartwell, J.L. 1976. Types of anticancer agents isolated from plants. Cancer Treat. Rep. 60:1031.

10. Wall, M.E. and M.C. Wani. 1977. Antineoplastic agents from plants. Annu. Rev. Pharmacol. Toxicol. 17:117.

11. Cordell, G.A. 1979. Anticancer agents from plants. Prog. Phytochem. 5:273.

12. Pettit, G.R. 1977. Biosynthetic products for cancer therapy. 1:228 pp.

13. Tomita, M., H. Furukawa, S.T. Lu and S.M. Kupchan. 1965. The constitution of thalicarpine. Tetrahedron Lett. 4309.

14. Kupchan, S.M., N. Yokoyama and B.S. Thyagarajan. 1961. Menispermaceae alkaloids. II. The alkaloids of Cyclea peltata Diels. J. Pharm. Sci. 50:164.

15. Kupchan, S.M. and A.J. Liepa. 1971. Total synethesis of the tumor-inhibitory alkaloids thalicarpine and hernandaline. Chem. Commun. 599.

16. Gilmore, C.J., R.F. Bryan and S.M. Kupchan. 1976. Conformation and reactivity of the macrocyclic tumor-inhibitory alkaloid tetrandrine. J. Am. Chem. Soc. 98:1947.

17. Sieber, S.M., J.A.R. Mead and R.H. Adamson. 1976. Pharmacology of antitumor agents from higher plants. Cancer Treat. Rep. 60:1127.

18. Herman, E.H. and D.P. Chadwick. 1974. Cardiovascular effects of d-tetrandrine. Pharmacology 12:97.

19. Panettiere, F. and C.A. Coltman, Jr. 1971. Phase I experience with emetine hydrochloride (NSC 33669) as an antitumor agent. Cancer 27:835.

20. Siddiqui, S., D. Firat and S. Olshin. 1973. Phase II study of emetine (NSC 33669) in the treatment of

solid tumors. Cancer Chemother. Rep. Part I, 57:423.

21. Wall, M.E., M.C. Wani, C.E. Cook, K.H. Palmer, A.T.
    McPhail and G.A. Sim. 1966. Plant antitumor agents.
    I. The isolation and structure of camptothecin, a
    novel alkaloidal leukemia and tumor inhibitor from
    Camptotheca acuminata. J. Am. Chem. Soc. 88:3888.

22. Wani, M.C. and M.E. Wall. 1969. Plant antitumor agents.
    II. The structure of two new alkaloids from
    Camptotheca acuminata. J. Org. Chem. 34: 1364.

23. Moertel, C.G., A.J. Schutt, R.J. Reitemeier and R.G.
    Hahn. 1972. Phase II study of camptothecin (NSC
    100880) in the treatment of advanced gastrointestinal
    cancer. Cancer Chemother. Rep. Part I, 56:95.

24. Gottlieb, J.A. and J.K. Luce. 1972. Treatment of
    malignant melanoma with camptothecin (NSC 100880).
    Cancer Chemother. Rep. Part 1, 56:103.

25. Schaeppi, U., R.W. Fleischman and D.A. Cooney. 1974.
    Toxicity of camptothecin (NSC 100880). Cancer
    Chemother. Rep. Part 3, 5:25.

26. Agarwal, J.S. and R.P. Rastogi. 1973. Chemical consti-
    tuents of Mappia foetida Miers. Indian J. Chem.
    11:969.

27. Tafur, S., J.D. Nelson, D.C. DeLong and G.H. Svoboda.
    1976. Antiviral components of Ophiorrhiza mungos.
    Isolation of camptothecin and 10-methoxycamptothecin.
    Lloydia 39:261.

28. Winterfeldt, E. 1975. Recent Dev. Chem. Nat. Carbon
    Compd. 6:9.

29. Shamma, M. and V. St. Georgiev. 1974. Camptothecin. J.
    Pharm. Sci. 63:163.

30. Schultz, A.G. 1973. Camptothecin. Chem. Rev. 73:385.

31. Svoboda, G.H., G.A. Poore and M.L. Montfort. 1968.
    Alkaloids of Ochrosia maculata Jacq. (Ochrosia
    borbonica Gmel.). J. Pharm. Sci. 57:1720.

32. Hayat, M., G. Mathé, M.M. Janot, P. Potier, N.
    Dat-Xuong, A. Cavé, T. Sevenet, C. Kan-Fan, J.
    Poisson, J. Miet, J. Le Men, F. LeGoffic, A.
    Gouyette, A. Ahond, L.K. Dalton and T.A. Connors.
    1974. Experimental screening of 3 forms and 19 deri-
    vatives or analogs of ellipticine: oncostatic
    effect on L1210 leukemia and immunosuppressive effect
    of 4 of them. Biomedicine 21:101.

33. Herman, E.H., D.P. Chadwick and R.M. Mhatre. 1974.
    Comparison of the acute hemolytic and cardiovascular
    actions of ellipticine (NSC 71795) and some ellipti-
    cine analogs. Cancer Chemother. Rep. Part 1, 58:637.

34. Svoboda, G.H., G.H. Poore, P.J. Simpson and G.B. Boder.
    1966. Alkaloids of Acronychia baueri Schott. I.
    Isolation of the alkaloids and a study of the anti-
    tumor and other biological properties of acronycine.
    J. Pharm. Sci. 55:758.

35. Svoboda, G.H. 1966. Alkaloids of Acronychia baueri
    (Bauerella australiana). II. Extraction of the
    alkaloids and studies of structure-activity
    relationships. Lloydia 29:206.

36. Schneider, J., E.L. Evans, E. Grunberg and R.I. Fryer.
    1972. Synthesis and biological activity of acronycine
    analogs. J. Med. Chem. 15:266.

37. Beck, J.R., R. Kwok, R.N. Booher, A.C. Brown, L.E.
    Patterson, P. Pranc, B. Rockey and A. Pohland. 1968.
    Synthesis of acronycine. J. Am. Chem. Soc. 90:4706.

38. Mulchandani, N.B. and S.R. Venkatachalam. 1976. Alka-
    loids of Pergularia pallida. Phytochemistry 15:1561.

39. Gellert, E., T.R. Govindachari, M.V. Lakshmikantham,
    I.S. Ragade, R. Rudzats and N. Viswanathan. 1962.
    The alkaloids of Tylophora crebriflora: structure and
    synthesis of tylocrebrine, a new phenanthroindolizi-
    dine alkaloid. J. Chem. Soc. 1008.

40. Farnsworth, N.R., A.S. Bingel, H.H.S. Fong, A.A. Saleh,
    G.M. Christenson and S.M. Sauffer. 1976. Oncogenic
    and tumor-promoting spermatophytes and pteridophytes
    and their active principles. Cancer Treat. Rep.60:1171.

41. Kupchan, S.M., R.W. Doskotch and P.W. Vanevenhoven. 1964. Tumor inhibitors. III. Monocrotaline, the active principle of Crotalaria spectabilis. J. Pharm. Sci. 53:343.

42. Kugelman, M., W. Liu, M. Axelrod, T.J. McBride and K.V. Rao. 1976. Indicine-N-Oxide: the antitumor principle of Heliotropium indicum. Lloydia 39:125.

43. Kupchan, S.M., Y. Aynehchi, J.M. Cassady, H.K. Schnoes and A.L. Burlingame. 1969. Tumor inhibitors. XL. The isolation and structural elucidation of elephantin and elephantopin, two novel sesquiterpenoid tumor inhibitors from Elephantopus elatus. J. Org. Chem. 34:3867.

44. Kupchan, S.M., R.J. Hemingway, D. Werner and A. Karim. 1969. Tumor Inhibitors. XLVI. Vernolepin, a novel sesquiterpene dilactone tumor inhibitor from Vernonia hymenolepis A. Rich. J. Org. Chem. 34:3903.

45. Kupchan, S.M., M.A. Eakin and A.M. Thomas. 1971. Tumor inhibitors. 69. Structure-cytotoxicity relationships among the sesquiterpene lactones. J. Med. Chem. 14:1147.

46. Rodriguez, E., G.H.N. Towers and J.C. Mitchell. 1976. Biological activities of sesquiterpene lactones. Phytochemistry 15:1573.

47. Mitchel, J.C. 1975. Contact allergy from plants. Recent Adv. Phytochem. 9:119.

48. Block, J.B., A.A. Serpick, W. Miller and P.H. Wiernik. 1974. Early clinical studies with lapachol (NSC 11905). Cancer Chemother. Rep. Part 2, 4:27.

49. Rao, K.V. 1974. Quinone natural products: streptonigrin (NSC 45383) and lapachol (NSC 11905) structure-activity relationships. Cancer Chemother. Rep. Part 2, 4:11.

50. Messmer, W.M., M. Tin-Wa, H.H.S. Fong, C. Bevelle, N.R. Farnsworth, D.J. Abraham and J. Trojánek. 1972. Fagaronine, a new tumor inhibitor isolated from Fagara zanthoxyloides Lam. (Rutaceae). J. Pharm. Sci. 61:1858.

51. Arthur, H.R., W.H. Hui and Y.L. Ng. 1959. An examina-
    tion of the rutaceae of Hong Kong. Part II. The
    alkaloids, nitidine and oxynitidine, from Zanthoxylum
    nitidum. J. Chem. Soc. 1840.

52. Gillespie, J.P., L.G. Amoros and F.R. Stermitz. 1974.
    Synthesis of fagaronine. An anticancer
    benzophenanthridine alkaloid. J. Org. Chem. 39:3239.

53. Ninomiya, I., T. Naito and H. Ishii. 1975. Synthesis of
    N-demethylfagaronine. Heterocycles 3:307.

54. Stermitz, F.R., K.A. Larson and D.K. Kim. 1973. Some
    structural relationships among cytotoxic and anti-
    tumor benzophenanthridine alkaloid derivatives. J.
    Med. Chem. 16:939.

55. Powell, R.G., D. Weisleder and C.R. Smith, Jr. 1972.
    Antitumor alkaloids from Cephalotaxus harringtonia:
    structure and activity. J. Pharm. Sci. 61:1227.

56. Corbett, T.H., D.P. Griswold, Jr., B.J. Roberts, J.C.
    Peckham and F.M. Schable, Jr. 1977. Evaluation of
    single agents and combinations of chemotherapeutic
    agents in mouse colon carcinomas. Cancer 40:2660.

57. Weinreb, S.M. and J. Auerbach. 1975. Total synthesis of
    the Cephalotaxus alkaloids. Cephalotaxine, cephalo-
    taxinone and demethylcephalotaxinone. J. Am. Chem.
    Soc. 97:2503.

58. Semmelhack, M.F., B.P. Chong, R.D. Stauffer, T.D.
    Rogerson, A. Chong and L.D. Jones. 1975. Total
    synthesis of the Cephalotaxus alkaloids. A problem
    in nucleophilic aromatic substitution. J. Am. Chem.
    Soc. 97:2507.

59. Mikolajczak, K.L., C.R. Smith, Jr., D. Weisleder, T.R.
    Kelly, J.C. McKenna and P.A. Christenson. 1974.
    Synthesis of deoxyharringtonine. Tetrahedron Lett.
    283.

60. Anon. 1975. Partial synthesis of harringtonine. K'o
    Hsueh T'ung Pao 20:437.

61. Mikolajczak, K.L. and C.R. Smith, Jr. 1978. Synthesis

of harringtonine, a Cephalotaxus antitumor alkaloid.
J. Org. Chem. 43:4762.

62. Kelly, T.R., R.W. McNutt, M. Montury, N.P. Tosches,
K.L. Mikolajczak, C.R. Smith, Jr. and D. Weisleder.
1979. Preparation of harringtonine from
cephalotaxine. J. Org. Chem. 44:63.

63. Mikolajczak, K.L. and C.R. Smith, Jr. 1979. Synthetic
cephalotaxine esters having antileukemic activity.
U.S. Patent 4,152,333.

64. Smith, C.R., Jr., K.L. Mikolajczak and R.G. Powell.
Harringtonine and related cephalotaxine esters. In
Design and Synthesis of Potential Anti-Cancer Agents
Based on Natural Product Models (J.M. Cassady and
J.D. Douros, eds). In press.

65. Anon. 1976. Cephalotaxine esters in the treatment of
acute leukemia, a preliminary clinical assessment.
Chin. Med. J. (Peking Engl. Ed.) 2:263.

66. Anon. 1977. Harringtonine in acute leukemias, clinical
analysis of 31 cases. Chin. Med. J. (Peking Engl.
Ed.) 3:319.

67. Wani, M.C., H.L. Taylor, M.E. Wall, P. Coggon and A.T.
McPhail. 1971. Plant antitumor agents. VI. The isolation and structure of taxol, a novel antileukemic and
antitumor agent from Taxus brevifolia. J. Am. Chem.
Soc. 93:2325.

68. Powell, R.G., R.W. Miller and C.R. Smith, Jr. 1979.
Cephalomannine; a new antitumor alkaloid from
Cephalotaxus mannii. Chem. Commun. 102.

69. Polonsky, J. 1973. Quassinoid bitter principles. Prog.
Chem. Org. Nat. Prod. 30:101.

70. Kupchan, S.M., R.W. Britton, J.A. Lacadie, M.F. Ziegler
and C.W. Sigel. 1975. The isolation and structural
elucidation of bruceantin and bruceantinol, new
potent antileukemic quassinoids from Brucea
antidysenterica. J. Org. Chem. 40:648.

71. Ogura, M., G.A. Cordell, A.D. Kinghorn and N.R.

Farnsworth. 1977. Potential anticancer agents. VI.
Constituents of Ailanthus excelsa (Simaroubaceae).
Lloydia 40:579.

72. Wani, M.C., H.L. Taylor, J.B. Thompson and M.E. Wall.
    1978. Plant antitumor agents. XVI. 6α-Senecioyloxy-
    chaparrinone, a new antileukemic quassinoid from
    Simaba multiflora. Lloydia 41:578.

73. Wall, M.E. and M.C. Wani. 1978. Plant antitumor agents.
    17. Structural requirements for antineoplastic acti-
    vity in quassinoids. J. Med. Chem. 21:1186.

74. Evans, F.J. and C.J. Soper. 1978. The tigliane,
    daphnane and ingenane diterpenes, their chemistry,
    distribution and biological activities. A review.
    Lloydia 41:193.

75. Kupchan, S.M., J.G. Sweeny, R.L. Baxter, T. Murae, V.A.
    Zimmerly and B.R. Sickles. 1975. Gnididin, gniditrin
    and gnidicin, novel potent antileukemic diterpenoid
    esters from Gnidia lamprantha. J. Am. Chem. Soc.
    97:672

76. Kupchan, S.M., Y. Shizuri, T. Murae, J.G. Sweeny, H.R.
    Haynes, M.S. Shen, J.C. Barrick, R.F. Bryan, D. van
    der Helm and K.K. Wu. 1976. Gnidimacrin and gnidi-
    macrin 20-palmitate, novel macrocyclic antileukemic
    diterpenoid esters from Gnidia subcordata. J. Am.
    Chem. Soc. 98:5719.

77. Kupchan, S.M. and R.L. Baxter. 1975. Mezerein: anti-
    leukemic principle isolated from Daphne mezereum L.
    Science 187:652.

78. Kupchan, S.M., Y. Komoda, W.A. Court, G.J. Thomas, R.M.
    Smith, A. Karim, C.J. Gilmore, R.C. Haltiwanger and
    R.F. Bryan. 1972. Maytansine, a novel antileukemic
    ansa macrolide from Maytenus ovatus. J. Am. Chem.
    Soc. 95:1354.

79. Wani, M.C., H.L. Taylor and M.E. Wall. 1973. Plant
    antitumor agents: colubrinol acetate and colubrinol,
    antileukaemic ansa macrolides from Colubrina
    texensis. Chem. Commun. 390.

80. Kupchan, S.M., Y. Komoda, A.R. Branfman, A.T. Sneden, W.A. Court, G.J. Thomas, H.P.J. Hintz, R.M. Smith, A. Karim, G.A. Howie, A.K. Verma, Y. Nagao, R.G. Dailey, Jr., V.A. Zimmerly and W.C. Sumner, Jr. 1977. The maytansinoids. Isolation, structural elucidation and chemical interrelation of novel ansa macrolides. J. Org. Chem. 42:2349.

81. Kupchan, S.M., A.T. Sneden, A.R. Branfman, G.A. Howie, L.I. Rebhun, W.E. McIvor, R.W. Wang and T.C. Schnaitman. 1978. Structural requirements for anti-leukemic activity among the naturally occurring and semisynthetic maytansinoids. J. Med. Chem. 21:31.

82. Chabner, B.A., A.S. Levine, B.L. Johnson and R.C. Young. 1978. Initial clinical trials of maytansine, an antitumor plant alkaloid. Cancer Treat. Rep. 62:429.

83. Kingsbury, J.M. 1964. Poisonous Plants of the United States and Canada, Prentice-Hall, pp. 353-357.

84. Powell, R.G., C.R. Smith, Jr. and R.V. Madrigal. 1976. Antitumor activity of Sesbania vesicaria, S. punicea and S. drummondii seed extracts. Planta Med. 30:1.

85. Powell, R.G., C.R. Smith, Jr., D. Weisleder, D.A. Muthard and J. Clardy. 1979. Sesbanine, a novel cyto-toxic alkaloid from Sesbania drummondii. J. Am. Chem. Soc. 101:2784.

86. Kupchan, S.M., W.A. Court, R.G. Dailey, Jr., C.J. Gilmore and R.F. Bryan. 1972. Triptolide and tripdiolide, novel antileukemic diterpenoid triepoxides from Tripterygium wilfordii. J. Am. Chem. Soc. 94:7194.

87. Kupchan, S.M., D.R. Streelman, B.B. Jarvis, R.G. Dailey, Jr. and A.T. Sneden. 1977. Isolation of potent new antileukemic trichothecenes from Baccharis megapotamica. J. Org. Chem. 42:4221.

88. Douros, J. and M. Suffness. 1978. New natural products of interest under development at the National Cancer Institute. Cancer Chemother. Pharmacol. 1:91.

89. Jolad, S.D., R.M. Wiedhopf and J.R. Cole. 1975. Tumor-inhibitory agent from Montezuma speciosissima (Malvaceae). J. Pharm. Sci. 64:1889.

90. Anon. February 19, 1979. Chem. Eng. News., p. 33.

91. Spencer, G.F., R.D. Plattner and R.G. Powell. 1976. Quantitative gas chromatography and gas chromatography-mass spectrometry of Cephalotaxus alkaloids. J. Chromatogr. 120:335.

92. Kupchan, S.M., C.W. Sigel, L.J. Guttman, R.J. Restivo and R.F. Bryan. 1972. Datiscoside, a novel anti-leukemic cucurbitacin glycoside from Datisca glomerata. J. Am. Chem. Soc. 94:1353.

93. Ogura, M., K. Koike, G.A. Cordell and N.R. Farnsworth. 1978. Potential anticancer agents. VIII. Constituents of Baliospermum montanum (Euphorbiaceae). Planta Med. 33:128.

94. Loder, J.W. and R.H. Nearn. 1975. 3β-Acetoxynorerythro-suamine, a highly cytotoxic alkaloid from Erythrophleum. Tetrahedron Lett. 2497.

95. Warthen, D., E.L. Gooden and M. Jacobson. 1969. Tumor inhibitors: liriodenine, a cytotoxic alkaloid from Annona glabra. J. Pharm. Sci. 58:637.

96. Galsky, A.G., J.P. Wilsey and R.G. Powell. 1980. Crown-gall tumor disc bioassay: a possible aid in the detection of compounds with antitumor activity. Plant Physiol. 65: 184.

97. Glogowski, W. and A.G. Galsky. 1978. Agrobacterium tumefaciens site attachment as a necessary pre-requisite for crown gall tumor formation on potato discs. Plant Physiol. 61:1031.

98. Richardson, C.L. and D.J. Morre. 1978. Chemotherapy of plant tumors: regression of crown gall tumors with D-glucosamine. Bot. Gaz. 139:196.

Chapter Three

SEARCH FOR CARCINOGENIC PRINCIPLES

JULIA F. MORTON

Morton Collectanea
University of Miami
Coral Gables, FL

Introduction
Carcinogens in the Flora of the Netherlands Antilles
Carcinogens in Other Parts of the World
Tea as a Possible Cause of Esophageal Cancer
The Effects of Wine as a Carcinogen
Correlation Between Tannin Content and Carcinogenesis
Other Compounds which may be Carcinogenic

INTRODUCTION

Worldwide, large-scale screening of plants for antitumor activity has been in progress for many years. The
testing of plants for carcinogenic or cocarcinogenic
factors proceeds at a lesser rate because it is the pre-
occupation of a relatively few investigators. Since plants
or plant products, crude or processed, form a major portion
of human and animal intake for dietary, recreational or
medical purposes, the evaluation of plants so consumed for
carcinogenic principles should receive high priority. Of
the total cancer deaths occurring annually in England and
Wales, 32% (approximately 40,000) are the result of tumors
in the esophagus, stomach, pancreas or large bowel.[12]
Esophageal cancer shows the most striking geographical
variation in the incidence of all cancers. This fact,
together with the high mortality rate, or usual short-term
survival rate, has aroused widespread interest in
causation. From my field surveys, observations and
research over a 15-year period in a search for possible

causes of esophageal cancer, there have evolved inescapable
correlations between this disease and high intakes of condensed catechin tannins and related anthocyanins through
plant materials deliberately ingested or entering the
system via occupational or other exposures.

CARCINOGENS IN THE FLORA OF THE NETHERLANDS ANTILLES

Let me review briefly the steps leading to these
correlations. In 1962, I was asked by the National Cancer
Institute to evaluate a list of plants ingested by the
Bantu people of the Transkei where there had been a
remarkable increase in esophageal cancer over the preceding
two decades. Then I was requested to compare this plant
list with the flora of the Netherlands Antilles because of
the reported high level of esophageal cancer on the island
of Curacao.[18] Impressed with the similarities in the vegetation of the two widely-separated areas, the National
Cancer Institute proposed an exploration of plant usage and
other practices and conditions in Curacao which might con-
ceivably relate to the incidence of the disease which, on
this island, affects men and women almost equally and has
fluctuated little over a period of 45 years.

I first went to Curacao to confer with Dr. Rolf
Eibergen, Pathologist, and the physicians comprising his
Working Group on Esophageal Cancer in the fall of 1964, and
began my field study in June of 1965. In my briefings at
Bethesda, I had been instructed to investigate the population make-up, the soil and climate, and all environmental
factors including the phosphate mining and the smoke from
the oil refinery. I had to conclude that these industries
were irrelevant because the refinery is close to the harbor
and the fumes blow offshore always in a southwesterly
direction. The phosphate mountain, Tafelberg, is at the
east end of Curacao. But the cases of esophageal cancer
are scattered the length of the island, though the majority
occur, of course, in and around the City of Willemstad
where most of the population is concentrated.

Water and food could be expected to be significant in
esophageal cancer but the water is almost entirely pure,
being distilled from seawater and carried by a slender pipe
to all inhabited rural districts. The island is burdened
with perennial drought and agriculture has diminished to a

minor status. The main agricultural crop and traditional
food is a brown-coated sorghum grown because it is drought-
resistant, and it is customarily eaten 3 times a day; as
porridge in the morning, as "funchi" at noon, and as
warmed-up "funchi" in the evening. Another local dish is
soup made from the dried, pulverized thin flesh of the
abundant, wild, columnar cactus, Cereus repandus, called
Cadushi. Some vegetables are raised by Portugese or
Chinese truck gardeners but most fresh produce is brought
from Venezuela or the Dominican Republic by sailboat and
sold in the picturesque "floating market" to those who can
afford to purchase such luxuries.

The ingestion of plant material arising from local
soils reaches significant proportions only in the consumption of sorghum and in the widespread and active use of
some native and exotic plants as "bush teas", or home
remedies. Folk-medicine plants became, therefore, prime
subjects for study. They had received no prior recognition
nor attention as potential factors in esophageal cancer in
Curacao.

Early steps in my assignment revealed that leaves,
roots and entire herbaceous plants are not only gathered
from the wild or from dooryards when needed but are also
regularly brought to the native market in Willemstad, fresh
or partially dried, and displayed in heaps and bundles. It
was a great aid to find that a local, elderly priest,
Father Paul Brenneker, had prepared a little book entitled
Jerba: Kruiden van Curacao en hun Gebruik (Herbs of
Curacao and their Uses), containing 500 Papiamento plant
names each followed by a few lines in Dutch relating medicinal uses as gleaned by questioning the natives. Father
Brenneker's acquaintance with the plants themselves is
limited.

It was necessary for me to work with the old and
current floras of the Netherlands Antilles (mostly in
Dutch) and with the plant vendors to establish the botanical identities corresponding to the colloquial names. It
became apparent that certain plants were dealt with under
several different names, and the 500 entries can be associated with approximately 104 species.

Most of these folk remedies are in common, current use
not only as remedies in illnesses, but some decoctions are

imbibed as daily beverages, as tonics or to "protect the
kidneys" after over-indulgence in alcohol. Some are taken
as aphrodisiacs or abortifacients and some routinely administered to infants just to induce sound sleep.[32] A popular cough sirup is made from the pulp of the wild calabash
(Crescentia cujete) after roasting the fruits and squeezing
out the juice. Among the most popular sources of "bush
teas" consumed by children and adults are: Mampuritu
(Porophyllum ruderale var. macrocephalum), Basora Pretu
(Cordia cylindrostachya), Sangura (Hyptis suavelolens),
Kleistubom (Passiflora foetida var. Moritziana), Oregano
(Lippia alba), Cocolode (Heliotropium angiospermum), Sali
(Heliotropium ternatum), leaves of Mango (Mangifera indica)
and Sorsaca (Annona muricata), Welensali (Croton flavens),
leaves and roots of Flaira (Jatropha gossypifolia), leaves,
stems and roots of Krameria ixina and Melochia tomentosa,
and roots of Acacia glauca (formerly known as A. villosa).
I obtained specimens from herb vendors and many private
individuals and collected plants in the wild for photo-
graphing and pressing and maintaining as vouchers in the
herbarium of the University of Miami.

In order to discover any possible association between
esophageal cancer and plant usage, and also to determine
prime suspects among the large number of plants involved, I
developed a questionnaire and interviewed hospitalized
patients, and also interrogated survivors of the deceased
in their hard-to-find homes in town and around the country-
side. I collected large quantities of the plants most commonly ingested by esophageal cancer victims for lyophili-
zing at the University of Miami and bioassay at the
National Cancer Institute.

Most extracts produced no significant lesions in
experimental animals. An extract of the leaves of Annona
muricata produced fibrosarcomas at the injection site in 5
out of 15 rats.[40] Pulverized leaves of this species in the
cheek pouch of hamsters induced in 2 of 16 animals super-
ficial spreading carcinoma, epithelial proliferation and
dyskeratosis, and basal cell proliferation extending down
into the esophagus. Pulverized, dried Heliotropium
ternatum, placed in hamster cheek pouches, produced epithelial atypia and loss of cell polarity in the esophagus of 4
animals; with calcium hydroxide added, 9 animals exhibited
papillomas in the upper third of the esophagus.[17] The
extract of Heliotropium angiospermum yielded sarcomas at

the injection site in 10 of 14 rats.[41] In 1969, among 22
plants in the testing program, 3 proved to be of extraor-
dinary interest. An extract of the supraterrestrial portion of Krameria ixina gave rise to fibrosarcomas at the
injection site in 100% of the rats, the first tumor
appearing after only 9 months. The root extract of Acacia
glauca also produced fibrosarcomas in 100% of the
recipients. The root extract of Melochia tomentosa caused
high mortality but after 1 : 10 dilution produced fibrosarcomas in all survivors.[39]

    While the bioassays at NIH were progressing, I under-
took a comparative study on the island of Aruba, only 45
miles from Curacao, with similar soil and climate, an oil
refinery (formerly two), but rarely any esophageal cancer.
I found that Krameria ixina and Acacia glauca do not exist
on Aruba. Melochia tomentosa occurs there, but the roots
are not used. Some people take the decoction of leafy
stems to halt diarrhea and, they say, to improve
circulation. Boiling water is poured on the flowers and
the infusion is drunk as a tonic. Some other plant uses
also vary from those of Curacao.[37] I found a different
kind of sorghum, white-seeded, called in Papiamento Maishi
di Shete Saman (7 week maize). This desirable grain has
never been successful in Curacao because that island is on
the migratory bird route and birds devour it avidly. Among
other dietary differences, I noted that while commercial
tea, without milk, is commonly consumed in Curacao by those
not economically limited to "bush tea", coffee is the principal adult beverage in Aruba. There is a much higher
intake of alcohol in Aruba than in Curacao.

    Later, because I had learned that people in Curacao
sometimes had Krameria ixina brought over from Venezuela by
friends or relatives, I ascertained the source as Coro, the
capital of the state of Falcon, just 35 miles away, and
made a survey of that area. There has been much cultural
interchange between Coro and Curacao for generations. The
native market of Coro resembles that of Curacao but is more
extensive. Many of the same folk-medicine plants are
conspicuous in the native vegetation and in the market.
These include Jatropha gossypifolia, Croton flavens,
Melochia tomentosa and Krameria ixina. Indians bring down
loads of Krameria from a nearby mountaintop where I saw it
growing profusely. I visited cancer clinics there and in
Maracaibo (the nearest large city) and compiled data

revealing a high rate of esophageal cancer in Coro in
contrast to the rest of the country.[34] After I published
these findings, the Instituto Venezolano de Investigaciones
Cientificas launched a year-long study, verified my data
and advised me that their survey revealed that a decoction
of Krameria ixina is habitually drunk by 30% of the population of the state.[29] Subsequently, I toured the native
markets across all of northern Venezuela from Caracas
westward, but I failed to see Krameria on sale anywhere and
gathered statistics attesting a low rate of esophageal
cancer throughout that vast area.[33]

    Pondering the carcinogenic potency of Krameria ixina,
Melochia tomentosa and Acacia glauca and the fact that the
three plants are totally unrelated botanically, I focused
on their similarity in all yielding an astringent, purple-
red decoction. Delving into old literature, I learned that
Dr. C. Hurtado, erstwhile Professor of Medicine in univer-
sities in Colombia, Venezuela and Argentina, had published
a compendium of the medicinal plants of Curacao in 1891-92
and stated that Krameria ixina was a good astringent and
hemostat with the same properties and uses as rhatany,
Krameria triandra, and could be substituted for that
imported drug. This led to the realization that all three
of our prime carcinogenic plants shared one feature in
common: a high level of condensed catechin tannins. Dr.
Rudi P. Labadie examined tops and roots of Melochia tomentosa sent to him at the Pharmaceutical Laboratory of the
University of Leiden, Netherlands, and advised me that the
roots contained 11.1% catechin tannin and the stems
6.1%.[27] I recalled that catechin tannin characterized the
betel nut (Areca catechu) which had long been associated
with oral and esophageal cancer in India and southern
Africa but suspicions rested on the additives in the betel
quid rather than on the nut itself. Recently, an extract
of betel nut alone has produced tumors at the injection
site in 100% of the animals in an experimental study at
Howard University. Dr. Govind Kapadia and his colleagues
there repeated subcutaneous injection of Krameria ixina
extract and obtained sarcomas in 80% of the rats. No
sarcomas developed from the injection of an aqueous extract
of the plant from which the tannin had been removed by
caffeine precipitation.[40]

CARCINOGENS IN OTHER PARTS OF THE WORLD

The development of my bar-graph showing at a glance the
world distribution of esophageal cancer (the low, red part
of each bar representing men and the upper, yellow, representing women) suggested the possibility of utilizing the
common denominator of catechin tannin to identify other
possible causes of the disease.

In 1970, I encountered the Sorghum Conversion Program
at Texas A & M and, through studying sorghum literature,
became aware of the high level of condensed catechin tannin
in dark, bird-, insect-, and fungus-resistant sorghums, and
also of the fact that a dark sorghum called Kaoliang is a
basic food of peasants in northern China where the incidence of esophageal cancer is claimed to be the highest in
the world.[38] The people complain of the bitter taste but
still grow this grain because of its high yield. In recent
years, the bird-resistant hybrids have been introduced from
Texas, and, while they may be somewhat more palatable, they
can hardly be expected to change the disease pattern.
Government officials are endeavoring to promote the culture
of maize in the northern Chinese provinces.[6] It may be
noteworthy that there is a progressive decrease in esophageal cancer rates as one goes south into the rice-growing
regions of China.

There are indigenous races of sorghum all over the con-
tinent of Africa, and the very light or white "Guinea corn"
strains are preferred throughout western Africa, but the
dark-brown or red-brown "Kaffir corn" strains of Kenya were
carried by the migrating Bantu to the Transkei and are preferred for Kaffir beer in many villages because they yield
a drink with the bitterness of hops. Also, wherever birds
are plentiful, the people are forced to grow the dark
types. In Kenya, white and light-brown varieties of
sorghum and finger millet are used for making porridge, but
dark-grained varieties for making beer. Esophageal cancer
is 14 times more prevalent among Luo tribesmen in western
Kenya who make their beer from dark sorghum and millet than
among their neighbors, the Kalenjins, who make their beer
from honey.[35,38]

TEA AS A POSSIBLE CAUSE OF ESOPHAGEAL CANCER

While analyzing epidemiological reports of many
investigators, I saw tea frequently mentioned, especially
in regard to the Cancer Belt of northern Iran, but only
because it is drunk while very hot. I quickly ascertained
that oxidised polymeric catechins which behave like
catechin tannins are an important element of tea and I
learned, by making specific inquiry, that tea is the main
remedy for diarrhea in northern Iran. For that purpose, it
has to be taken strong! While the percentage of tannin in
tea varies considerably with the strain grown, the country
of origin, the stage of growth (that is, the age of the
leaves plucked), the degree of oxidative fermentation, the
blend and the brand, it may be as high as 23% in certain
Formosa Oolongs. Some mild Chinese and Japanese green teas
may possess as little as 4%, quite unlike black teas preferred in Ceylon, Indonesia, India, Holland, Great Britain
and North America, which may contain 19 to 22%. Black
instant tea produced in Russia is extolled as containing
all the catechins of the tea leaf. It includes 27 to 33%
of tannins, 91.5 to 147.7 mg./g. of catechins.[42]

In an earlier paper,[35] I quoted from tea trade
literature, in part, as follows: "Tannin, in any part of
the alimentary canal, the mouth, the gullet, the stomach or
the intestines, may be harmful, depending upon the quantity
taken, their other contents and their constitution. When
milk is added to tea, the casein therein fixes the tea
tannin, and prevents its action on the mucous membrane of
the mouth, and on that part of the alimentary canal leading
to the stomach ... Many efforts have been made to remove or
reduce the tannin in tea ... The tannin products were
removed by filtration and the remaining liquor, tannin-
free, was claimed to have no injurious effect on the
digestive organs ... none of the processes for making tea
free from tannin have made headway. The fact is that the
removal of tannin decreases not only the astringency, but
also the strength and body of the tea, and the physiologic
effect takes second place to taste, in the tea drinker's
mind." Thanks to early warnings by the British Medical
Association, the British have always put milk in their tea.
The Dutch don't; and about 100 years ago esophageal cancer
was rife in Holland. When the Dutch switched from tea to
coffee as the national drink, the disease became rare.

Following my tannin/cancer links, Dr. Mitsuo Segi, head of the Segi Institute of Cancer Epidemiology in Japan, made a survey of the high-risk provinces for esophageal cancer in his country and established a direct correlation between the disease and the habitual consumption of tea-rice-gruel.[44] I have recently been informed of an American man who took up tea drinking while in the Army in Korea in the 1950's because coffee was not always available. Returning home, he became an electrician and took a large thermos of tea without milk to work every day. During the winter, he would also drink tea at home in the evening. In 1974, he succumbed of esophageal cancer at the age of 46.[24]

I have interviewed the widow of a lawyer who started drinking iced tea daily when he came to the warm Miami climate in 1949. The tea was prepared in advance, kept in the refrigerator, and served, not with shaved ice, but with ice cubes which did not melt readily in the cold liquid and therefore caused little dilution. My questions elicited the fact that there were always ice cubes left in the glass when it was refilled. The gentleman drank 3 or 4 or more glasses every day, without added milk. He never used tobacco, indulged little in spirits, but his dentist had commented on his stained teeth. He developed esophageal cancer and expired at the age of 53.[14] I am often asked if the addition of lemon affects the tannin content. It does nothing but flavor the tea.

THE EFFECTS OF WINE AS A CARCINOGEN

Because my bar-graph showed a general proliferation of esophageal cancer all over the wine-producing regions of western Europe, I was encouraged to look into wine and, thanks to the excellent texts, Phenolic Substances in Grapes and Wine, and their Significance,[45] Technology of Wine Making,[1] and other literature, I quickly learned that dry red wine is rich in condensed catechin tannins, and my investigation of the case of a prominent Dutch clergyman of Curacao, who died of esophageal cancer, revealed that he drank every evening a very dry Burgundy. I found that a Portugese farmer who also succumbed to the disease had regularly received from his mother in Madeira shipments of red wine made from very small grapes. Since tannin is highest in the seeds of grapes, secondarily in the skin, the smaller the grapes the more tannin in the vintage. In

62   J. F. MORTON

a study of 226 cases of cancer of the mouth and pharynx in
Switzerland, investigators found that persons consuming 5
deciliters or more of wine daily accounted for 82% of the
patients. The incidence was highest among smokers who were
also heavy drinkers.[22]

I am greatly concerned about the strong promotion of
grape and wine production in the United States. Extensive
plantings are being made of the cultivars of high-tannin
native species. It is gleefully anticipated that wine consumption in the United States will approach 400 million
gallons by 1980, a 60% increase over the 1970 level.[5] In
the past, the wine industry of California was largely based
on the European grape, Vitis vinifera, which, because of
its low-phenol content is highly disease-susceptible
elsewhere. It was the exporting of high-phenol, disease-
resistant native American grape vines as rootstocks that
saved the European wine growers from extinction by
Phylloxera root-rot in the 1890's, and which also raised
the tannin level in European wines. I heard last summer of
a heavy demand by California wine makers for Muscadine
grape juice from South Carolina to flavor the wine. This
will certainly elevate the tannin level. Incidentally,
deer graze on Vitis vinifera but not on our native V.
labrusca,[46] just as goats in Curacao will not touch non-
spiny Acacia glauca, but struggle to reach the tiny
leaflets of Acacia tortuosa which nature has armed with
vicious thorns instead of phenols.

Dr. Hans Breider, Director of the Bavarian National
Institute for Viticulture, Fruit-growing, and Horticulture,
Wurzburg, Bavaria, has expressed great concern over the
"possible consequences in men and animals of the chemico-
physiological effect of chemical elements of the resistance
of the vine to its parasites." In his experiments with the
wine from direct-producing, disease-resistant hybrids,
malformations and liver damage were demonstrated in chicks,
and alcohol was eliminated as a causative factor.[9] Similar
deformities are being observed presently at Purdue
University in chicks fed the bird-, insect-, and disease-
resistant, high-tannin hybrid sorghums, and attempts are
being made to detoxify these sorghums.[11]

Researchers at the Bureau of Microbial Hazards of the
Canadian Department of Health and Welfare announced in 1976
that poliovirus, herpes simplex virus, and various enteric

viruses are inactivated by incubation with wine or diluted
wine. Red wine is much more effective than white wine
because of its higher phenol content. Plain red grape
juice is even more active having a phenol content of 3.3 g.
per liter as compared with 2.2 g. per liter in red wine
and 0.26 g. per liter in white wine.[25]

Of course, aging in oak casks, or quick-aging with oak
chips, enhances the tannin content of wines and other alcoholic beverages. Ethanol is colorless. Few drinkers stop
to ask themselves where the amber color of Scotch, Bourbon
or rum comes from. Aging is sometimes accelerated by
heating. Puerto Rican chemists have recently reported success in quick-aging rum in 18 weeks at 115° F. in place of
the traditional storage for many months or years. Color
was enhanced and, in 140° Proof rum, tannin concentration
was elevated by 39%.[47]

There have been many disquisitions on alcohol and
cancer, simply because many victims of gastrointestinal
cancer have been heavy drinkers, but, if alcohol per se
were carcinogenic, esophageal and stomach cancer would be
common throughout the United States, inasmuch as alcoholism
is one of our greatest social problems. Actually, esophageal cancer is steadily increasing in this country only
among black males and they are known to be the leading consumers of cheap red wine. The French, trying to link alcohol and esophageal cancer, found the strongest correlation
in the Normandy Peninsula, but I have pointed out that, in
that region, too far north for grapes, they make and drink
dry cider from high-tannin, inedible apples and pears.[35]

CORRELATIONS BETWEEN TANNIN CONTENT AND CARCINOGENESIS

After I had reported to the National Cancer Institute
my apparent correlations of tannin ingestion and esophageal
cancer, I was asked to choose any high-risk area for this
disease in the United States and begin a new investigation
to test the evidence. Being aware of reports of an unusual
frequency of esophageal cancer cases seen by surgeons in
the hospitals of Charleston, South Carolina, I undertook a
survey of the Charleston area and surrounding Low Country.
I examined and extracted data from the records of 894 cases
covering a span of 32 years -- 1940 through 1971.
Distribution of the 888 cases for which addresses were

found revealed 3 coastal counties and 5 adjoining inland
counties as markedly exceeding the state rate. These 8
counties constitute the Low Country where the people have,
since the Civil War, continued the practice of self-
medication with local plants.

Gathering information directly from the people and
collecting plant materials, I uncovered a remarkable intake
of tannin in "bush-teas" taken primarily as remedies for
diarrhea, dysentery, colds, influenza, sorethroats and
hemorrhages. These included decoctions of the bark of the
cherrybark oak (Quercus falcata var. pagodaefolia), leafy
branches of wax myrtle (Myrica cerifera), root of the marsh
rosemary (Limonium nashii). They drink frequently as a
cold remedy and as a beverage a decoction of the young
shoots and needles at the branch tips of the longleaf pine
(Pinus palustris). Women take this "tea" regularly to
relieve menstrual cramps. They are overly fond of the
bitter, astringent decoction of life-everlasting
(Gnaphalium obtusifolium), sometimes flavored with sugar or
lemon extract. They drink it as a cold remedy and febri-
fuge alone, and often boil the plant with wax myrtle and
"pine tops" for a more potent dose. The tannin-rich leaves
and petioles of sweet gum (Liquidambar styraciflua) are
chewed to relieve sorethroat and halt diarrhea. The root
of the trailing blackberry (Rubus trivialis) is chewed to
stop diarrhea. In 1869, Dr. Francis Peyre Porcher
declared: "I have known cases of chronic diarrhea and
dysentery which recovered after using a strong tea of
blackberry root, which had resisted other and persistent
efforts for their relief." Blackberry wine is a standard
remedy for diarrhea in South Carolina as well as being a
very popular home-made beverage.[36] According to Peter
Valaer, "Blackberry secretes much more tannin than does any
other wine, a fact that accounts for its medical uses."[35]

I have furnished quantities of a number of the Low
Country folk-remedy plants to researchers at Howard
University and, in bioassay, an aqueous extract of the bark
of cherrybark oak produced tumors in 100% of the rats
receiving the injection; the tannin fraction induced tumors
in 28 of 30 animals. Aqueous extracts and tannin fractions
of wax myrtle, roots of marsh rosemary, and sweet gum
leaves also proved to be highly carcinogenic.[23]

The phenolic intake of the Low Country people is
enhanced by much imbibing of elderberry wine and grape
wine, by the utilization of the thick skins and often a few
seeds of the Muscadine grape in a traditional grape pie,
the chewing of sections of purple-stemmed sugarcane, and
drinking the raw juice of unpeeled stems direct from the
farm mill; also by chewing tobacco and, in the case of
women, using snuff orally, depositing it behind the lower
lip and rubbing it on the gums.

In an occupational analysis of Low Country esophageal
cancer victims, I discovered an extra risk among men
working as carpenters and in sawmills and attributed this
to the swallowing of sawdust.[36] This view was strengthened
by my comparative analysis of occupations of 789 prostatic
and 519 cervical cancer cases, and by reports of an excess
of nasopharyngeal or ethmoid tumors among woodworkers in
England, France and Denmark.[2,13] In 1977, there appeared
the first announcement of unusually high rates of nasal
cancer among furniture makers in North Carolina.[10] And
adenocarcinoma of the nose and paranasal sinuses in Swedish
men has been linked to exposure to wood dust.[19] These
revelations are long past due.

People are the best experimental animals. We need to
know how they successfully produce cancer in themselves.
We should inquire into this while the patient is still
alive. But I have seen, in my reviews of cancer cases,
that the most neglected part of the patient's admission
record is the irritant history. In my examination of
esophageal cancer files in South Carolina, I found no mention whatsoever of the use of botanical folk remedies and
no reference to wine. In only 2 of the 1,308 files on
prostatic and cervical cases were there notations as to
herb tea and watermelon seed tea used by a male and his
father, and the drinking of life-everlasting and pine tops
by a female.[36] Cancer Registry forms are of little help
when they list occupations such as "bookkeeper", "cashier",
and so on with no reference to the industry. When a
patient is no longer employed, a simple check mark in the
blank space after the word "Retired" is usually all that
one finds.

If more attention had been paid to patients' habits,
cancer and other diseases would not have remained so long
such mysteries. Fortunately, with the environmental

consciousness of the past 10 years, there is greater emphasis on causes rather than nearly total concentration on
therapy. Epidemiologists should broaden their concepts and
not go forth seeking to establish correlations with one or
two preconceived carcinogenic factors such as aflatoxins
and nitrosamines, which may safely be assumed to be
universal. They should examine all aspects of the human
condition. When adequate information is effectively
assembled and sorted out by computer, it will reveal
patterns of disease and their causal relationships. Then,
hopefully, educational and preventive measures may follow.
Unfortunately, there are now strict governmental regulations designed to restrict and hamper the gathering of such
essential information.

We all consume potential carcinogens daily. It is
important to point out that the risk of ingesting tannin-
containing plant products is in proportion to the quantity,
frequency and length of time. For example, esophageal
cancer is highest among Asiatic women in Africa who chew
betel nut all day long, and low among men who chew betel
only when they come home from work at night. The reason
that esophageal cancer is most common among low-income
populations is that they have a limited choice of foods and
beverages, and therefore ingest these excessively; and,
also, in their folk remedies they favor those with obvious
immediate results -- astringent, tannin-rich plant
materials shrinking tissues in inflamed throats, halting
diarrhea, dysentery and hemorrhages.

It is interesting to note that high-tannin plants like
yellow dock (Rumex crispus) and sweet pond lily (Nymphaea
odorata) were old-time remedies for cancer in the South;
and that tannin, like radiation, can work both ways -- if
tumors are present, cause regression; can, in excess, over
a period of time, cause tumors in healthy tissue.[20] Perdue
and Hartwell reported in 1969 that of 240 plants showing
antitumor action, the activity in 35% was due to tannins.[41]

Another effect of tannins is interference with the
utilization of protein. In Florida, steers fed non-bird-
resistant (low tannin) sorghum gained weight 16.3% faster
and were 19.6% more efficient in converting feed to gain
than those receiving dark, bird-resistant (DeKalb BR-64)
sorghum.[7] This deficit in nutritional value cancels out a
large part of the 25 to 30% higher yield per acre of the BR

hybrids. Strains of the broadbean (Vicia faba) from white-
flowered plants with tannin-free seedcoats have a 4 to 7%
higher digestibility than strains with colored flowers and
tannin-containing seedcoats. The digestibility of the
tannin-free seedcoats is 56.4% compared with 17.2% for
tannin-containing seedcoats.[8] In Costa Rica, where white
types of the common bean (Phaseolus vulgaris) are preferred
over colored types, experiments were conducted with a view
to developing non-black mutants of the locally high yielding black bean cultivars. A white-seeded mutant, NEP-2,
resulting from gamma radiation, was selected as having the
most desirable agronomic and food characteristics. In
feeding experiments by the Instituto de Nutricion de
Centroamerica y Panama (INCAP), it has doubled the weight
gain produced by its black progenitor in a 24-day
period.[4,31]

There is no question that disease- and pest-resistance
and high yields in beans, as in grapevines and sorghum, are
related to the phenolic content of the growing plant. The
commercial demand for white-seeded snapbeans is based on
the standard of clear liquor in the processed product.
Breeders have expressed the opinion that this "eye-appeal"
quality requirement places an unfair burden on breeders and
growers who are striving to improve bean productivity.[15]
However, the greatest consideration should be given to the
dietary aspects of cultivar color, governed by phenolic
content. With the current great pressures on agricultur-
ists to produce more food per acre and reduce the use of
pesticides we need active intercommunication betwen plant
breeders, nutritionists and cancer researchers in order to
avoid costly, misdirected programs which, measured in terms
of effectively feeding, not merely filling, the hungry, are
self-defeating, and which may increase rather than lower
the levels of carcinogens in the human diet.

OTHER COMPOUNDS WHICH MAY BE CARCINOGENIC

The risk of exposure to carcinogens may vary confus-
ingly through concurrent exposure to carcinostats or cocarcinogens. Jatropha gossypifolia, which I have mentioned,
and which is much employed as a folk remedy in Curacao,
Coro and Aruba, has shown considerable tumor-inhibiting
activity in laboratory experiments[26] though it belongs to

the family Euphorbiaceae, many species of which have tumor-
promoting properties.

There is increasing interest in cocarcinogens, and the
first Co-Carcinogenesis Symposium, sponsored by the
National Cancer Institute, was held at the Oak Ridge
National Laboratory in March 1977. Professor Dr. Erich
Hecker, Director of the Biochemical Institute of the German
Center for Cancer Research in Heidelberg, who specializes
in a study of cocarcinogens in the Euphorbiaceae, is
intensely interested in Croton flavens which I referred to
above as a common source of "bush tea" in Curacao and Coro.
From plant material which I have repeatedly shipped to him
from Curacao, he has isolated phorbol esters which he has
demonstrated to be potent cocarcinogens.[48] I also prepared
and shipped to him 40 liters of the "tea". On July 10th of
this year (1979) he notified me that his assistant had isolated from this decoction a di-terpene triester fraction
which is similarly active as an irritant.[21,28]

According to a recent release from the Wistar
Institute, phorbol esters are as ambidextrous as tannins,
being able to inhibit or stimulate cell differentiation
along normal or alternate pathways.[43] This dual activity
is apparently common in the plant world. A team of
investigators in the Departments of Surgery and
Physiological Chemistry at Ohio State University have just
released a report that withdrawal of caffeine and other
methylxanthines (discontinuance of coffee, tea, chocolate
and cola drinks) brought about complete disappearance of
all palpable breast nodules, pain, tenderness and nipple
discharge in 13 of 20 women with fibrocystic disease and
other benign breast lesions within 1 to 6 months. In no
patient was there a progression from benign lesions to
cancer.[30]   Caffeine, at low levels, has demonstrated protective action in animals exposed to known carcinogens.[3]
There has even been some hopeful speculation that the pre-
valence of caffeine in the particulate organic matter in
the air of New York City (arising from coffee-roasting
plants there and in New Jersey) might offset the cancer
hazard from the polynuclear aromatic hydrocarbons and other
carcinogens in the air pollution.[16] Nevertheless, at
extremely high levels, caffeine is mutagenic and
teratogenic.[3]

REFERENCES

1. Amerine, M.A., H.W. Berg and H.V. Cruess. 1967. The Technology of Wine Making. The AVI Pub'g Co., Westport, Conn. 799 pp.

2. Andersen, H.C. 1975. Exogenic causes of cancer of the nose (Orig. in Danish). Ugeskr Laegr 137(44): 2567-2571. (Carcinogenesis Abs. 14(12):7115, 1976.)

3. Anonymous. 1976. Caffeine, coffee and cancer. Brit. Med. J. 1(6017):1031-1032.

4. Anonymous. 1975. El desarollo de variedades de frijol para llenar los requisitos latinamericanos. La Hacienda 70(6):22-23.

5. Anonymous. 1970. News of the western fruit industry. Blue Anchor 47(4):29-30.

6. Anonymous. 1975. Plant Studies in the People's Republic of China: A Trip Report of the American Plant Studies Delegation. Nat'l Acad. Sci., Washington, D.C. 206 pp.

7. Bertrand, J.E. and M.C. Lutrick. 1971. The feeding value of bird-resistant and non-bird-resistant sorghum grain in the ration of beef steers. Sunshine State Agric. Res. Rpt. 16(3):16-17.

8. Bond, D.A. 1976. In vitro digestibility of the testa in tannin-free field beans (Vicia faba L.). J. Agric. Sci., Camb. 86:561-566.

9. Breider, H. (undated). Quality and resistance of the vine. I.N. Investigaciones Agronomicas 28(59):289-307. Reilly Translations, Gardena, Calif.

10. Brinton, L.A., W.J. Blot, B.J. Stone and J.F. Fraumeni, Jr. 1977. A death certificate analysis of nasal cancer among furniture workers in North Carolina. Cancer Res. 37:3473-3474.

11. Butler, Dr. L., Dept. Biochemistry, Purdue Univ.,
    West Lafayette, Ind. Pers. Comm. Feb. 12, 1979.

12. Cummings, J.H. 1978. Dietary factors in the aetiology
    of gastrointestinal cancer. J. Hum. Nutr.
    32(16):455-465. (Abs. HEW, ICRDB Cancergram Ser.
    CK03, #5; 1979).

13. Curtes, J.P., E. Trotel and J. Bourdiniere. 1977.
    Adenocarcinomas of the ethmoid region in woodworkers.
    Arch. Mal. Prof. 38(9):773-786. (Abs. HEW, ICRDB
    Cancergram Ser. CK02, #5; 1978).

14. Davison, V. Pers. Comm., May 22, 1978.

15. Deakin, J.R. and P.D. Dukes. 1975. Breeding snap beans
    for resistance to diseases caused by Rhizoctonia
    solani Kuehn. Hort. Sci. 10(3):269-271.

16. Doug, M., D. Hoffman, D.C. Locke and E. Ferrand. 1977.
    The occurrence of caffeine in the air of New York
    City. Atmos. Environ. 11(7):651-653.

17. Dunham, L.J., R.H. Sheets and J.F. Morton. 1974.
    Proliferative lesions in cheek pouch and esophagus of
    hamsters treated with plants from Curacao,
    Netherlands Antilles. J. Nat'l Cancer Inst. 53(5):
    1259-1269.

18. Eibergen, R. 1961. Kanker op Curacao. Groningen. J.B.
    Wolters. 135 pp.

19. Engzell, U., A. Englund and P. Westerholm. 1978. Nasal
    cancer associated with occupational exposure to orga-
    nic dust. Acta Otolaryngol. 86(5/6):437-442. (Abs.
    HEW, ICRDB Cancergram Ser. CK02, #5; 1979.)

20. Farnsworth, N.R., A.S. Bingel, H.H.S. Fong, A.A. Saleh,
    G.M. Christenson and S.M. Saufferer. 1976. Oncogenic
    and tumor-promoting spermatophytes and pteridophytes
    and their active principles. Cancer Treatment Rpts.
    60(8):1171-1214.

21. Hecker, E., Direktor, Institut fur Biochemie, Deutsches
    Krebsforschungszentrum, Heidelberg, Germany. Pers.
    comm., July 10, 1979.

22. Junod, B. and R. Pasche. 1978. Etiology and epidemi-
    ology of cancer of the mouth and pharynx in
    Switzerland. Schweiz Med. Wochenschr. 108(24):
    882-887. (HEW, ICRDB Cancergram Ser. CK02, #4; 1979.)

23. Kapadia, G.J., B.D. Paul, E.B. Chung, B. Ghosh and S.N.
    Pradhan. 1976. Carcinogenicity of Camellia sinensis
    (tea) and some tannin-containing folk medicinal herbs
    administered subcutaneously in rats. J. Nat'l Cancer
    Inst. 57(1):207-209.

24. Knowles, H.A., Washington, D.C. Pers. comm., June 27,
    1979.

25. Konowalchuk, J. and J.I. Speirs. 1976. Virus inactiva-
    tion by grapes and wines. J. Appl. & Environ.
    Microbiol. 32(6):757-763.

26. Kupchan, S.M., C.W. Sigel, M.J. Matz, A.S. Renauld,
    R.C. Haltiwanger and R.F. Bryan. 1970. Jatrophone, a
    novel macrocyclic diterpenoid tumor inhibitor from
    Jatropha gossypifolia. J. Am. Chem. Soc.
    92:4476-4477.

27. Labadie, R.P., Farmaceutisch Laboratorium van de
    Rijksuniversiteit te Leiden. Pers. comm., May 24,
    1971.

28. Lutz, D. and E. Hecker. 1979. Esophageal cancer on
    Curacao - further new tumor promoters from Croton
    flavens L. Abs. of paper presented at the 8th
    International Symp. on the Biol. Characterisation
    of Human Tumours, Athens, Greece, May 8-11, 1979.

29. Merino, Dr. F. Departamento de Medicina Experimental,
    Instituto Venezolano de Investigaciones Cientificas,
    Caracas, Venezuela. Pers. comm., April 22, 1977.

30. Minton, J.P., M.K. Foecking, D.J.T. Webster and R.H.
    Matthews. 1979. Response of fibrocystic disease to
    caffeine withdrawal and cyclic nucleotides correlated
    with breast disease. Amer. J. Obstet. (in press).

31. Moh, C.C. 1974. The white bean mutant (Phaseolus
    vulgaris), p. 18. In: The application of nuclear
    energy to agriculture. COO-3217-26. Trop. Agric.
    Res. & Training Center, Turrialba, Costa Rica.

32. Morton, J.F. 1968. A survey of medicinal plants of Curacao. Econ. Bot. 22(1):87-102.

33. Morton, J.F. 1975. Current folk remedies of northern Venezuela. Quart. J. Crude Drug Res. 13:97-121.

34. Morton, J.F. 1974. Folk-remedy plants and esophageal cancer in Coro, Venezuela. Morris Arbor. Bull., Univ. of Penna. 25:24-34.

35. Morton, J.F. 1972. Further associations of plant tannins and human cancer. Quart. J. Crude Drug Res. 12(1):1829-1841.

36. Morton, J.F. 1973. Plant products and occupational materials ingested by esophageal cancer victims in South Carolina. Quart. J. Crude Drug Res. 13(1): 2005-2022.

37. Morton, J.F. 1968. Plants associated with esophageal cancer cases in Curacao. Cancer Res. 28:2268-2271.

38. Morton, J.F. 1970. Tentative correlations of plant usage and esophageal cancer zones. Econ. Bot. 24(2): 217-226.

39. O'Gara, R.W., C.W. Lee and J.F. Morton. 1971. Carcinogenicity of extracts of selected plants from Curacao after oral and subcutaneous administration to rodents. J. Nat'l Cancer Inst. 46: 1131-1137.

40. O'Gara, R.W., C.W. Lee, J.F. Morton, G.J. Kapadia and L.J. Dunham. 1974. Sarcoma induced in rats from extracts of plants and from fractionated extracts of Krameria ixina. J. Nat'l Cancer Inst. 52(2):445-448.

41. Perdue, R.E. and J.L. Hartwell. 1969. The search for plant sources of anticancer drugs. Morris Arbor. Bull. 20(3):35-53.

42. Pruidze, G.N., D.V. Gogisvanidze and M.A. Bokuchava. 1968. A comparative study of chemical properties of dry concentrates of black instant tea. (Orig. in Russian). Priklad Biokhim. Mickrobiol. 4(4):478-480. (Biol. Abs. 50:18274; 1969).

43. Rovera, G., T.G. O'Brien and L. Diamond. 1979. Induction of differentiation in human promyelocytic leukemia cells by tumor promoters. Science 204:868-870.

44. Segi, M. 1975. Tea-gruel as a possible factor for cancer of the esophagus. Gann 66: 199-202.

45. Singleton, V.L. and P. Esau. 1969. Phenolic Substances in Grapes and Wine and Their Significance. Academic Press, New York. 182 pp.

46. Teranishi, R., U.S. Dept. of Agric., Western Regional Res. Lab., Albany, Calif. Pers. comm., Aug. 28, 1975.

47. Torres, C.S., J.L. Aguiar, H. Batiz and I. Hernandez. 1979. Aging of rum in a hot chamber. J. Agric. Univ. Puerto Rico 63(1):64-77.

48. Weber, J. and E. Hecker. 1978. Cocarcinogens of the diterpene ester type from Croton flavens and esophageal cancer in Curacao. Experientia 34:679-682.

Chapter Four

GLYCOALKALOIDS OF THE SOLANACEAE

STANLEY F. OSMAN

Eastern Regional Research Center
U.S. Department of Agriculture
Philadelphia, Pennsylvania   19118

Introduction
The Chemistry of the Glycoalkaloids
The Biosynthesis of Glycoalkaloids
Glycoalkaloid Distribution in the Plant
Biological Activity
Toxicity of Glycoalkaloids
Conclusion

INTRODUCTION

   Glycoalkaloids are nitrogenous steroidal glycosides
that are found in most Solanum species. Defosses, in
1820,[1] reported that the active principle of morel (S.
nigrum) was an organic base which he named solanine.
Baup[2] reported the presence of solanine in potatoes and
concluded it "will find a use in medicine..." More than a
century later, Solanum glycoalkaloids have become important
starting compounds for the commercial preparation of
steroidal hormone intermediates. Glycoalkaloid research
has not been restricted to chemical studies; the biological
activity of these compounds has been extensively investigated, primarily in the context of plant resistance to
pests and microorganisms and of human toxicity. The presence of glycoalkaloids in foods such as potatoes,
tomatoes, and eggplant has always been of great concern;
understandably, this concern has generated much research
activity.

TABLE 1
COMPOSITION OF SOLANUM GLYCOALKALOIDS

I   R = H
I_a R = H, Δ^5 UNSAT.
I_b R = H Δ^5 UNSAT., 23 $O-\overset{O}{\overset{\|}{C}}-CH_3$

II   R = H
II_a R = H, Δ^5 UNSAT.

Glycoalkaloid	Aglycone	Aglycone Structure	R Carbohydrate
α-Solanine	Solanidine	I_a (20S,22R,25S)	Solatriose
α-Chaconine	Solanidine	I_a (20S,22R,25S)	Chacotriose
Solasonine	Solasodine	II_a (22R,25R)	Solatriose
Solamargine	Solasodine	II_a (22R,25R)	Chacotriose
Tomatine	Tomatidine	II (22S,25S)	Lycotetraose
Solacauline	Soladulcidine	II (22R,25R)	Polyatriose
α-Solamarine	Tomatidenol	II_a (22S,25S)	Solatriose
β-Solamarine	Tomatidenol	II_a (22S,25S)	Chacotriose
Demissine	Demissidine	I (20S,22R,25S)	Lycotetraose
Leptine I	Acetylleptinidine	I_b (20S,22R,25S)	Chacotriose
Leptine II	Acetylleptinidine	I_b (20S,22R,25S)	Solatriose
Commersonine	Demissidine	I (20S,22R,25S)	Commertetraose

Excellent reviews of the chemistry of glycoalkaloids include those by Prelog and Jeger[3,4] and Schreiber.[5] There is also a comprehensive review of tomatine chemistry and biology[6] and one of the glycoalkaloid literature emphasizing health-related research.[7] In this chapter many aspects of the chemistry, biochemistry and biology of the glycoalkaloids are discussed, with emphasis on research reported within the last ten years.

THE CHEMISTRY OF THE GLYCOALKALOIDS

The glycoalkaloid composition of more than 250 Solanum species has been determined.[5] The structures of many of these compounds are listed in Tables 1 and 2. The aglycone skeletal structure is either of the solanidane (I) or spirosolane (II) type. Minor structural variations of these two ring systems such as Δ^5 unsaturation or isometrization at C-22 account for most of the other aglycones listed in Table 1. The carbohydrate moieties listed in Table 2 can be found in combination with a number of different aglycones; for example, chacotriose is found in α-chaconine (solanidine aglycone), β-solamarine (tomatidenol aglycone) and solamargine (solasodine aglycone). Thus the multiplicity of glycoalkaloids found in Solanum species stems from minor modification of the aglycone structure and various combinations of aglycone and carbohydrate moieties. The compilation in Tables 1 and 2 is not meant to be complete; minor glycoalkaloids and compounds that may be derived from higher saccharides by hydrolysis (the removal of a rhamnose from α-chaconine yields β-chaconine) have been omitted. In the latter case, these compounds are artifacts produced by the action of hydrolytic enzymes released when the tissue is excised from the plant. If the enzymatic activity of S. tuberosum detached flowers is not immediately quenched, copious amounts of β-chaconine are produced; however, when precautions are taken to destroy post harvest enzymatic activity, β-chaconine is isolated only in minor amounts and α-chaconine is the predominant glycoalkaloid. Therefore, unless the proper precautions are taken, the origin of glycoalkaloids that may be derived from other glycoalkaloids by loss of sugar(s) is uncertain.[5]

It was not until the 1940's that characterization of Solanum glycoalkaloids was undertaken. The outstanding research of Prelog, Kuhn and Schreiber has contributed

TABLE 2

STRUCTURE OF GLYCOALKALOID CARBOHYDRATE MOIETIES

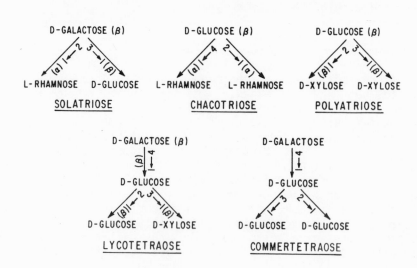

Figure 1. Glycoalkaloids

significantly to our knowledge of glycoalkaloid structures.
Relatively few new glycoalkaloids have been discovered in
the last ten years. The commercial importance of solasodine (or tomatidenol) as a precursor to the steroidal hormone intermediate 3β-acetoxy-pregna-5,16-diene-20-one has
been the primary incentive for continued characterization
of the glycoalkaloid composition of the Solanum species.
Solasodine glycoalkaloids recently isolated and characterized are listed in Table 3. Solatifoline was shown to be
different than solasonine (Table 1) by x-ray analysis,
although it contains the same aglycone and sugars.

New glycoalkaloids, containing the novel aglycone
congestidine (III), isolated from S. congestiflorum and
characterized by Katz et al.[13] are: solacongestinine
(III$_a$), α-solacongestinine (III$_b$), and β-solacongestinine
(III$_c$) as shown in Figure 1.

Modern methods of structural analysis have greatly
facilitated glycoalkaloid characterization. Radeglia et
al.[14] and Weston et al.[15] have reported the ^{13}C-NMR
spectra of tomatidine, solasodine, soladulcidine,
solanidine, and demissidine. With few exceptions, the
shift assignments are unambiguous. The structure of the
carbohydrate moiety can now be determined by a modified
permethylation method. The methylated sugars resulting
from hydrolysis of the permethylated glycoalkaloid are
characterized by combined gas chromatography - mass
spectrometry of the alditol acetate derivatives[16] rather
than the older classical methods[17] which required substantially larger quantities for identification.

Recently effort has been devoted to developing methods
for total glycoalkaloid (TGA) analysis because of the
suspected toxic activity of these compounds. Zitnak[7]
reviewed many of the available TGA methods. Most of these
methods are a compromise, since it is difficult to isolate,
in quantitative yield, a glycoalkaloid fraction that is
suitable for accurate analysis by either colorimetric or
gravimetric methods. Some of these methods also require
the presence of functional groups that are not common to
all glycoalkaloids such as olefinic unsaturation for the
formaldehyde-sulfuric acid reaction.[18] A method based on
titration of the free aglycone has been described;[19,20]
this method does not require extensive preliminary purification of the glycoalkaloid fraction and will measure all

TABLE 3

NEW SOLASODINE GLYCOALKALOIDS

Glycoalkaloid	Carbohydrate[a]	Solanum species	Reference
Solardixine	$-gal \xrightarrow{3-1} glu \xrightarrow{2-1} glu \xrightarrow{2-1} rham$	S. lactiniatum S. khasianum	Bite et al.[8]
Solasurine	$-glu \xrightarrow{} rham$	S. elaeagnifolium S. aviculare	Seth and Chatterjee[9]
Solashabanine	gal,3 glu, rham	S. lactiniatum	Bite and Shabana[10]
Solaradinine	gal,4 glu, rham	S. lactiniatum	Bite and Shabana[10]
Solapersine	gal, glu, 2 xyl	S. persicum	Novuzov et al.[11]
Solatifoline	glu, gal, rham	S. platanifolium	Puri and Bhatmagar[12]

[a] glu = glucose, gal = galactose, rham = rhamnose, xyl = xylose

glycoalkaloids. Rapid procedures are available for quanti-
tating specific glycoalkaloids. Solanum tuberosum tuber
glycoalkaloids (α-solanine and α-chaconine) can be determined satisfactorily by reaction of antimony trichloride[21]
or formaldehyde-sulfuric acid[22] with the base precipitated
glycoalkaloids followed by colorimetric analysis. A rapid,
semiquantitative analysis of potato glycoalkaloids using
thin-layer chromatography (TLC) has recently been
described.[23]

Qualitative glycoalkaloid analysis is accomplished by
either gas chromatography[24] or by TLC.[25] Unambiguous glycoalkaloid identification generally can be made by a combination of GC and TLC analysis.

Glycoalkaloids that differ by only minor modifications
in the aglycone structure (e.g., solanidine or demissidine
glycoalkaloids containing the same carbohydrate differ only
by the degree of saturation between C-5 and C-6) may
require more sophisticated techniques for separation such
as argentation chromatography. A method has been described
for distinguishing Δ^5 unsaturated glycoalkaloids from the
corresponding saturated compounds based on their behavior
to acid hydrolysis.[26a]

BIOSYNTHESIS OF GLYCOALKALOIDS

The biosynthesis of the aglycones of Solanum species
has been studied extensively; however, all the steps in the
pathway have not been delineated. Biosynthesis proceeds
from acetyl-CoA via the usual intermediates through
cholesterol[26b,27] to the various aglycones. The pathway
from cholesterol to the aglycones has been partially
inferred by the close relationship between steroid alkaloids and sapogenins in a given species, particularly with
respect to the configuration at C-25. A hypothetical pathway for the formation of the solanidane, congestidine and
spirosolane aglycones is shown in Figure 2. Although dor-
mantinol and dormantinone have been isolated from plants
that synthesize solanidine,[28] they have not been confirmed
as intermediates by labeling experiments. The interconver-
sion at C-22 has been postulated by Tschesche and
Spindler.[29] The nitrogen apparently is derived from
arginine.[30]

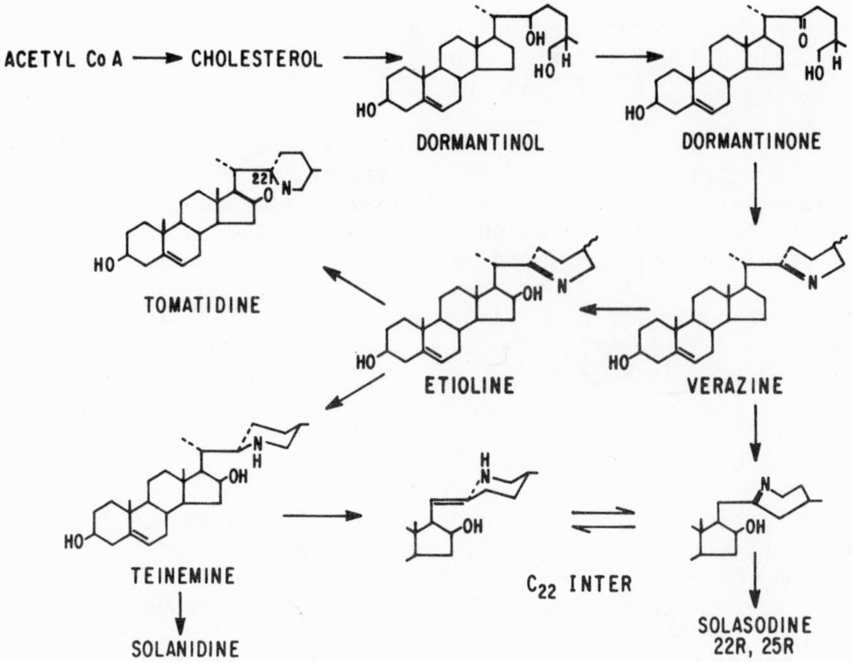

Figure 2.   Aglycone biosynthetic pathway

TABLE 4

DISTRIBUTION OF GLYCOALKALOIDS IN POTATO
AND TOMATO PLANT

Plant organ	% Glycoalkaloid (in dry tissue) Potato[a]	Tomato[b]
Flower	1.6 - 3.5[c]	0.9 - 2.0
Leaf	0.5 - 0.6	.5 - 5.1
Stem	.03 - .06	.08 - 0.6
Sprouts	0.6 - 4.1	-
Root	0.1	0.2 - 0.6
Tuber	.006 - .04	-
Fruit	-	.087 - .036[d]

a   Lampitt et al.[36]
b   Roddick[6]
c   Values include free solanidine
d   Range decrease in levels during fruit ripening which
    fall to undetectable levels after 2-3 days beyond ripe
    (red) stage.

TABLE 5

GLYCOALKALOID DISTRIBUTION IN POTATO TUBER

(from Lampitt et al.[36])

Tuber part	Part % of tuber	Total glycoalkaloids mg/100 dry wt
Skin	2	64
Peel (skin & outer cortex)	11	68 - 75
Flesh	90	21

The glycosylation of solanidine[31,32] and solasodine[33] has been demonstrated, although it has not been established whether the biosynthesis of the glycoalkaloid from the aglycone occurs through a step-wise addition of sugars similar to flavone glycoside synthesis.[34] Exogenous solanidine is metabolized in a step-wise manner to yield a diglucosyl solanidine,[35] which may be indicative of the endogenous glycosylation process.

GLYCOALKALOID DISTRIBUTION IN THE PLANT

Glycoalkaloids are usually found in all organs of the plant; the flowers contain the highest concentration of glycoalkaloids. In Table 4 the concentrations found in different organs of potato plants are compared with those reported for tomato. The values reported for potato were determined by Lampitt et al.[36] using the method of Rooke et al.;[37] the ranges of values reflect varietal and environmental effects on glycoalkaloid concentrations. The wide range of glycoalkaloid levels in tomatoes may also result from these effects and/or different methods of glycoalkaloid analysis. The distribution of glycoalkaloids in the potato tuber is shown in Table 5.

Glycoalkaloid concentration tends to be high in regions of high metabolic activity, such as meristems[38] and sprouts.[32,36] Synthesis takes place predominantly in plant tops.[39] In both tomato[38] and potato,[40] grafting experiments have shown that glycoalkaloid synthesis in the shoot (leaves and stem) is independent of synthesis in the root. Tomato[38,41] and potato[36] plant tissues metabolize glycoalkaloids and presumably change glycoalkaloid distribution within the plant as it matures. Glycoalkaloid metabolites have not been characterized; however, Sander[42] suggested that tomatine may be utilized in lycopene synthesis. In the maturing potato plant, glycoalkaloid concentration increases in the flowers, stolon, and tubers while decreasing in other plant organs.[36]

The intracellular distribution of tomatine has been determined by Roddick.[43] Organelles of pericarp tissue of green tomatoes were fractionated by centrifugation, and tomatine was found in the 105,000 g supernatant with a small amount in the microsomal fraction. Expressed juice of the fruit was also high in tomatine. These results

TABLE 6

TGA OF TUBERS FROM SELECTED SOLANUM SPECIES

Species	β-Chaconine	α-Chaconine	Glycoalkaloid α-Solanine	Solamarines[a]	Demissine	Tomatine
S. ajanhuiri[b]	3.5[c]	39.0	57.3			
S. curtilobum		34.8	46.4	5.3	13.4	
S. stenotonum	5.5	69.8	24.7			
S. juzepczukii		14.0	37.8	7.7	40.4	
S. acaule 1[d]					95.5	
2					62.1	30.9
3					88.2	11.6
4					64	34

a   Combined value for α- and β-solamarine.
b   All species are cultivated except for S. acaule.
c   Values represent percent of total glycoalkaloids.
d   Four clones of species S. acaule were analyzed.

suggest that, in contrast to sterols which are found mainly in membrane fractions, glycoalkaloids accumulate in vacuoles and/or the soluble phase of the cytoplasm. Synthesis may occur in the microsomal organelles.

Genetic and environmental factors determine the TGA levels found in a particular plant. In Table 6, the TGA values for tubers of some tuber-bearing Solanum species are tabulated; these species are considered to have potential use as breeding stock.[44] The intraspecies as well as the interspecies variation is evident. Intraspecies variation is probably due to environmental factors. TGA levels increase with increased exposure to light,[32] other factors such as temperature and soil condition may affect TGA to a lesser extent, but there is little data from which any correlations can be drawn. The cumulative effect of environment on TGA in a number of commercial varieties has been examined.[45,46] Results for these varieties similar to those shown in Table 6 demonstrate that environmental factors can significantly alter the TGA levels. Analysis of these studies reveals that a "low TGA" variety (Irish Cobbler, average TGA = 6.2 mg/100 g FW) when grown in Alaska was higher in TGA level than a "high TGA" variety (Kennebec, average TGA = 9.7) grown in Texas; viz 10.9 vs 5.8.

Mechanical damage to potatoes not only increases TGA levels[47,48] but may also cause qualitative changes in glycoalkaloid composition, such as the formation of α- and β-solamarine in slices of Kennebec tubers.[49] Damaged potatoes that are found in the market do not contain excessively high TGA levels.[50]

BIOLOGICAL ACTIVITY

The function of glycoalkaloids in the plant is a controversial issue. Hegnauer[51] coined the term "alcaloida imperfecta" to describe glycoalkaloids because he considered these compounds to be nitrogen derivatives of steroids and their nitrogen content and basicity accidental rather than essential characteristics. Fraenkel[52] suggested a role for secondary metabolites, in general, based partly on Schreiber's[53] data for the pest repellant activity of Solanum glycoalkaloids. The finding that the species S. chacoense, which is resistant to the Colorado potato

beetle,[54] has high TGA levels and also contains the leptine
glycoalkaloids,[55] which are highly repellant to the beetle
in feeding tests,[56] rekindled interest in glycoalkaloids as
significant factors for natural plant resistance to pests.
In field experiments with S. chacoense x S. tuberosum
hybrids, Schwarze was not able to correlate TGA levels with
resistance to Colorado potato beetle.[57] Prior to the
characterization of the leptines, Schreiber[53] hypothesized
that the necessary structure for maximum repellancy to the
Colorado potato beetle was a saturated aglycone with a
tetrasaccharide unit containing xylose. This hypothesis
was based on glycoalkaloid feeding experiments. However,
the leptines, which were subsequently recognized as probably the most active known glycoalkaloids toward the
beetle do not have either of these structural features.

Tingey et al.[58] and Raman et al.[59] have presented con-
vincing evidence that glycoalkaloids confer Solanum species
resistance to the potato leafhopper (Empoasca fabae
[Harris]) both in field experiments ($r = -0.75$, $P = 0.01$)
and in feeding studies ($r = -0.86$, $P = 0.01$). These
results support the earlier conclusions of Dahlman and
Hibbs.[60]

Although glycoalkaloids are toxic to many microorganisms, there appears to be no correlation between plant
glycoalkaloid level and field resistance. Deahl et al.[61]
were not able to establish any relationship between glycoalkaloid levels and field resistance to late blight in 15
potato clones; similarly, Langcake et al.[62a] found no
correlation between glycoalkaloid levels in tomato roots
and stems and resistance to Fusarium oxysporum f.
lycopersici. However, Mohanakumain and coworkers found[62b]
that varieties of Lycopepsicon pimpinellifolium resistant
to Pseudomonas solanacearum had higher root tomatine levels
than susceptible varieties.

Pathogens may have the ability to detoxify glycoalkaloids. Septoria lycopersici (leaf spot fungus of tomato)
hydrolyzes tomatine to the trisaccharide β_2-tomatine.[63]
Solanidine was produced in minor amounts when solanine was
incubated with P. infestans.[64]

Glycoalkaloids may have a role as general, nonspecific
protective agents against microbial or insect invasion.
Tomatine has been shown to be effective against dermato-

mycetes particularly Trichophyton mentagrophytes[65] and
fungotoxic levels of glycoalkaloids to the nonpathogen of
potatoes, Helminthosporium carbonum, have been found in
potato peels.[66] Arneson and Durbin suggested that high,
localized tomatine concentration may inhibit fungal
growth.[63] Although tomatine localization would not be
reflected in TGA levels for the whole plant, this hypothesis has not received experimental confirmation. Even in
the correlation of potato resistance to leafhopper[58]
mentioned above, the authors realize that factors other
than glycoalkaloid levels may be involved in plant
resistance. Generally, the level of glycoalkaloids
necessary to impart significant pest resistance would be
too high to be considered as an acceptable means of
resistance in food and feed crops (see below); however,
these compounds may serve such a function in wild species
in which they are present in considerable quantity.

TOXICITY OF GLYCOALKALOIDS

     Glycoalkaloids have been suspected as the causative
agent in potato poisoning even though there is little
evidence in the literature to substantiate this claim.
Zitnak[7] has presented an excellent review of this subject
from which one may conclude that the amount of glycoalkaloids lethal to humans is still an unanswered question;
however, these compounds are responsible for undesirable
effects such as vomiting, nausea, and diarrhea. The animal
toxicity of pure glycoalkaloids has been determined only
within the last 10 years. The LD_{50} (oral, >1000 mg/kg) reflects the low absorption through the intestinal wall into
the blood stream; 80% of radioactive α-chalonine is excreted
within 48 hours.[69] Nishie et al.[70] determined the cardio-
tonic activity of six glycoalkaloids and one aglycone; the
activity, in part, was correlatable to the number of sugars
attached to the aglycone.

     Renwick[71] suggested that potato glycoalkaloids may be
teratogens. Animal studies with pure glycoalkaloids have
been negative;[68,72,73] however, Keeler et al.[74] reported
significant birth defects in the offspring of golden
hamster females on diets containing potato sprout extracts.

CONCLUSIONS

Glycoalkaloids are still considered important starting compounds for steroidal hormone synthesis. The production of these compounds in cultured <u>Solanum</u> tissue is presently being investigated.[75] The only edible tissue that contains significant amounts of glycoalkaloids is potato tuber. The composition and concentration of glycoalkaloids found in commercial potatoes are acceptable at present; however, the use of new species should be carefully monitored for glyco-alkaloids throughout the breeding program. The variety Lenape, which was derived, in part, from <u>S. chacoense</u> (a generally high TGA species containing other glycoalkaloids in addition to the normal tuber glycoalkaloids, α-solanine and α-chaconine), was withdrawn from production because of its high TGA content.[1] Research on the genetics of glyco-alkaloid inheritance has been reported;[76,77] and further studies of this type are necessary to determine to what extent glycoalkaloid inheritability is predictable. Improvement of the natural resistance of plants to pests and microorganisms has become an important goal today because of the environmental consequences of indiscriminate use of pesticides. Glycoalkaloids may contribute to natural resistance, but the toxicity of these compounds and the lack of knowledge of the efficacy of glycoalkaloids as natural pesticides mitigates against breeding programs designed to alter glycoalkaloid composition and content of plants at the present time. Continued research in pest resistance and heredity of glycoalkaloids in edible plants, particularly the potato, may make it possible to incorpor-ate glycoalkaloids as part of a pest resistance program in the future.

REFERENCES

1. Zitnak, A. and G.R. Johnston. 1970. Glycoalkaloid
     content of B5141-6 potatoes. <u>Amer</u>. <u>Pot</u>. <u>J</u>. <u>47</u>:256.

2. Baup, M. 1826. Concerning several new substances. <u>Ann</u>.
     <u>Chim</u>. <u>31</u>:108.

3. Prelog, V. and O. Jeger. 1953. Steroid alkaloids: the
     <u>Solanum</u> group. In: <u>The Alkaloids</u>, (R.H.F. Manske,
     ed.) Academic Press (New York), Vol. III, p. 248.

4. Prelog, V. and O. Jeger. Ibid., 1960. Vol. VII, p. 343.

5. Schreiber, K. 1968. Steroid alkaloids: the Solanum
   group. In: The Alkaloids, (R.H.F. Manske, ed.)
   Academic Press (New York), Vol. X, p. 1.

6. Roddick, J.G. 1974. The steroidal glycoalkaloid
   α-tomatine. Phytochemistry 13:9.

7. Zitnak, A. 1977. Steroids and capsaicinoids of sola-
   naceous food plants. In: The Nightshades and Health,
   (N.F. Childers and G.M. Russo, eds.) Norman F.
   Childers, publisher. Somerset Press, Inc.
   (Somerville, NJ), p. 41.

8. Bite, P., M.M. Shabana, L. Jokay and L. Pomgraczne
   Sterk. 1969. Solanum glycosides IV. Solaradixene.
   Magy. Kem. Foly. 75:544.

9. Seth, D.K. and R. Chatterjee. 1968. Solasodine
   glycoalkaloids from Solanum species. J. Inst. Chem.
   (Calcutta) 41:194.

10. Bite, P. and M.M. Shabana. 1972. Solanum type
    glycosides VIII. Solashabanine and solaradinine. Acta
    Chim. (Budapest) 73:361.

11. Novuzov, E.N., S.M. Aslanov, N.M. Isonailov and A.A.
    Imanova. 1975. Glycoalkaloid from Solanum persicum.
    Khun Prir. Soedin 11:434.

12. Puri, R.K. and J.K. Bhatmagar. 1975. Solatifoline.
    Phytochemistry 14:2096.

13. Katz, R., N. Aimi and Y. Sato. 1975. New steroidal
    glycoalkaloids from congestiflorum. Adv. Front. Plant
    Sci. 30:63.

14. Radeglia, R., G. Adam and H. Ripperger. 1977. ^{13}C NMR
    spectroscopy of Solanum steroid alkaloids. Tet. Lett.
    903.

15. Weston, P.J., H.E. Gottlieb, E.W. Hagaman and E.
    Wenkert. 1977. Carbon-13 nuclear magnetic resonance
    spectroscopy of naturally occurring substances. LI.
    Solanum glycoalkaloids. Aust. J. Chem. 30:917.

16. Bjorndal, H., C.G. Hellerqvist, B. Lindberg and S. Svensson. 1970. Gas liquid chromatography and mass spectrometry in methylation analysis of poly-saccharides. Angew. Chem. (int'l. edit.) 9:610.

17. Kuhn, R., I. Low and H. Trisckmann. 1955. The constitution of solanine. Chem. Ber. 88:1492.

18. Schwarze, P. 1962. Methods for identification and determination of solanine in potato breeding material. Zuchter 32:155.

19. Fitzpatrick, T.J. and S.F. Osman. 1974. A comprehensive method for the determination of total potato glyco-alkaloids. Amer. Pot. J. 51:318.

20. Fitzpatrick, T.J., J.D. MacKenzie and P. Gregory. 1978a. Modifications of the comprehensive method for total glycoalkaloid determination. Amer. Pot. J. 55:247.

21. Bretzloff, C.W. 1971. A method for the rapid estimation of glycoalkaloids in potato tubers. Amer. Pot. J. 48:158.

22. Gull, D.P. and T.M. Isenberg. 1960. Chlorophyll and solanine content and distribution in four varieties of potato tubers. Proc. Amer. Soc. Hort. Sci. 75:545.

23. Cadle, L.S., D.A. Stelzig, K.L. Hayser and R.J. Young. 1978. Thin-layer chromatographic system for identification and quantitation of potato tuber glycoalkaloids. J. Agric. Food Chem. 26:1453.

24. Herb, S.F., T.S. Fitzpatrick and S.F. Osman. 1975. Separation of potato glycoalkaloids by gas chromatography. J. Agric. Food Chem. 23:520.

25. Boll, P.M. 1962. Alkaloid glycosides from Solanum dulcamara. Acta. Chem. Scand. 16:1819.

26a Osman, S.F. and S.L. Sinden. 1977. Analysis of mixtures of solanidine and demissidine glycoalkaloids contain-ing identical carbohydrate units. J. Agric. Food Chem. 25:955.

26b Heftmann, E., E.R. Lieber and R.D. Bennett. 1967.
    Biosynthesis of tomatidine from cholesterol in
    Lycopersicon pimpinellifolium. Phytochemistry 6:225.

27. Tschesche, R. and H. Hulpke. 1967. Biosynthesis of
    steroid derivatives in plants. VIII. Biogenesis of
    solanidine from cholesterol. Z. Naturforsch. 22:791.

28. Kaneko, K., M.W. Tanaka and H. Mitsuhashi. 1977.
    Dormantinol, a possible precursor in solanidine
    biosynthesis from budding Veratrum grandiflorum.
    Phytochemistry 16:1247.

29. Tschesche, R. and M. Spindler. 1978. Zur biogenese des
    aza-oxa-spiran-systems der steroidalkaloide vom
    spirosolan-typ in solanaceen. Phytochemistry 17:251.

30. Kaneko, K., M.W. Tanaka and H. Mitsuhashi. 1976.
    Origin of nitrogen in the biosynthesis of solanidine
    by Veratrum grandiflorum. Phytochemistry 15:1391.

31. Lavintman, N., J. Tandecary and C.E. Cardini. 1977.
    Enzymatic glycosylation of steroid alkaloids in
    potato tuber. Plant Sci. Letters 8:65.

32. Jadhav, S.J. and D.K. Salunke. 1975. Formation and
    control of chlorophyll and glycoalkaloids in tubers
    of Solanum tuberosum L. and evaluation of glycoalka-
    loid toxicity. In: Advances in Food Research, Vol.
    21, 307 (C.O. Chichester, ed.), Academic Press (New
    York).

33. Liljegren, D.R. 1971. Glycosylation of solasodine by
    extracts from Solanum lactiniatum. Phytochemistry
    10:3061.

34. Harborne, J.B. 1963. Plant polyphenols IX. The glyco-
    sidic pattern of anthocyanin pigments. Phytochemistry
    2:85.

35. Osman, S.F. and R.M. Zacharius. 1979. Biosynthesis of
    potato glycoalkaloids abstract. 63rd Annual Meeting
    of the Potato Assoc. of America, Vancouver, B.C.,
    July 12-15.

36. Lampitt, L.H., J.H. Bushill, H.S. Rooke and E.M. Jackson. 1943. Solanine, glycoside of the potato (II) distribution in the plant. J. Soc. Chem. Ind. 62:48.

37. Rooke, H.S., J.H. Bushill, L.H. Lampitt and E.M. Jackson. 1943. Solanine, glycoside of potato. Its isolation and determination. J. Chem. Soc. Ind. 62:20.

38. Sander, H. 1956. Studies on the formation and degradation of tomatine in the tomato plant. Planta 47:374.

39. Schreiber, K. 1954. The glycoalkaloids of solanaceae. Chem. Technik 6:648.

40. Prohoshev, S.M., E.I. Petrochenko, G.S. Ilin, V.Z. Baranova and N.A. Lebedeva. 1952. Glycoalkaloids in leaves and tubers of vegetatively grafted nightshade plants. Doklody Akad. Nauk. SSSR 83:881.

41. Wolf, M.J. and B.M. Duggar. 1940. Solanine in the potato and the effects of some factors on its synthesis and distribution. Amer. J. Bot. 27, Suppl. 20s.

42. Sander, H. 1958. Tomatine, a possible starting material for lycopene synthesis. Naturwissenschaften 45:59.

43. Roddick, J.G. 1976. Intracellular distribution of the steroidal glycoalkaloid α-tomatine in Lycopersicon esculentum fruit. Phytochemistry 15:475.

44. Osman, S.F., S.F. Herb, T.J. Fitzpatrick and P. Schmiediche. 1978. Glycoalkaloid composition of wild and cultivated tuber-bearing Solanum species. J. Agric. Food Chem. 26:1246.

45. Sinden, S.L. and R.E. Webb. 1972. Effect of variety and location on the glycoalkaloid content of potatoes. Amer. Pot. J. 49:334.

46. Sinden, S.L. and R.E. Webb. 1974. Effect of environment on glycoalkaloid content of six potato varieties at 39 locations. Tech. Bull. No. 1472. Agric. Research Service, U.S. Dept. of Agriculture.

47. Parups, E.V. and I. Hoffman. 1967. Induced alkaloid levels in potato tubers. Amer. Pot. J. 44:277.

48. Fitzpatrick, T.J., S.F. Herb, S.F. Osman and J.A. McDermott. 1977. Potato glycoalkaloids: increases and variations of ratios in aged slices over prolonged storage. Amer. Pot. J. 54:539

49. Shih, M. and J. Kuc. 1974. α- and β-solamarine in Kennebec Solanum tuberosum leaves and aged tuber slices. Phytochemistry 13:997.

50. Fitzpatrick, T.J., J.A. McDermott and S.F. Osman. 1978b. Evaluation of injured commercial potato samples for total glycoalkaloid content. J. Food Sci. 43:1617.

51. Hegnauer, R. 1964. The Chemotaxonomy of Plants, Vol. 3. Burkhauser, Basel.

52. Fraenkel, G.S. 1958. The raison d'etre of secondary plant substances. Science 129:1466.

53. Schreiber, K. 1957. Natural resistance factors of plants against the potato beetle and their possible mode of action. Zuchter 27:289.

54. Schaper, Von P. 1953. The resistance of S. chacoense to the Colorado potato beetle. Zuchter 23:115.

55. Kuhn, R. and I. Low. 1961. The constitution of the leptines. Chem. Ber. 94:1088.

56. Sturckow, B. and I. Low. 1961. Effect of some solanaceous alkaloidal glycosides on the potato beetle. Entomol. Exp. Appl. 4:133.

57. Schwarze, P. 1963. On the glycoalkaloid content and composition of the glycoalkaloid complex in the progeny Solanum tuberosum x Solanum chacoense crosses. Zuchter 33:275.

58. Tingey, W.M., J.D. Mackenzie and P. Gregory. 1978. Total foliar glycoalkaloids and resistance of wild potato species to Empoasca fabae (Harris). Amer. Pot. J. 55:577.

59. Raman, K.V., W.M. Tingey and P. Gregory. 1979. Potato glycoalkaloids: effect on survival feeding behavior of the potato leafhopper. J. Econ. Entomol. 72:337.

60. Dahlman, D.L. and E.T. Hibbs. 1967. Responses of Empoasca fabae (Cicadellidae:Homoptera) to tomatine, solanine, leptine I, tomatidine, solanidine and demissidine. Ann. Entomol. Soc. Am. 60:732.

61. Deahl, K.L., R.J. Young and S.L. Sinden. 1973. A study of the relationship of late blight resistance to glycoalkaloid content in fifteen potato clones. Amer. Pot. J. 50:248.

62a Langcake, P., R.B. Drysdale and H. Smith. 1972. Post-infectional production of an inhibitor of Fusarium oxysporum f. lycopersici by tomato plants. Physiol. Plant Path 2:17.

62b Mohanakumaran, N., J.C. Gilbert and I.W. Buddenhagen. 1969. Relationship between tomatine and bacterial wilt resistance in tomato. Phytopathology 59:14.

63. Arneson P.A. and R.D. Durbin. 1967. Hydrolysis of tomatine by Septoria lycopersici, a detoxification mechanism. Phytopathology 57:1358.

64. Holland, H.L. and G.J. Taylor. 1979. Transformations of steroids and the steroidal alkaloid, solanine, by Phytophthora infestans. Phytochemistry 18:437.

65. Tschesche, R. and G. Wolff. 1965. Antimicrobial action of saponins. Z. Naturforsch 20b:543.

66. Allen, E.H. and J. Kuc. 1968. α-Solanine and α-chaconine as fungitoxic compounds in extracts of Irish potato tubers. Phytopathology 58:776.

67. Nishie, K., M.R. Cushmann and A.C. Keyl. 1971. Pharmacology of solanine. Toxicol. Appl. Pharmacol. 19:81.

68. Nishie, K., W.P. Norred and A.P. Swain. 1975. Pharmacology and toxicology of chaconine and tomatine. Res. Commun. Chem. Path. α Pharmacol. 12:657.

69. Norred, W.P., K. Nishie and S.F. Osman. 1976.
    Excretion, distribution and metabolic fate of ^{3}H-α-
    chaconine. Res. Commun. Chem. Path. Pharmacol.
    13:161.

70. Nishie, K., T.J. Fitzpatrick, A.P. Swain and A.C. Keyl.
    1976. Positive ionotropic action of Solanaceae glyco-
    alkaloids. Res. Commun. Chem. Path. Pharmacol 15:601.

71. Renwick, J.H. 1972. Hypothesis. Anencephaly and spina
    bifida are usually preventable by avoidance of a spe-
    cific but unidentified substance present in certain
    potato tubers. Brit. J. Prev. Soc. Med. 26:67.

72. Mun, A.M., E.S. Barden, J.M. Wilson and J.M. Hogan.
    1975. Teratogenic effects in early chick embryos of
    solanine and glycoalkaloids from potato infected with
    late-blight, Phytophthora infestans. Teratology
    11:73.

73. Swingard, C.A. and S. Chaube. 1973. Are potatoes tera-
    togenic for experimental animalsπ Teratology 8:349.

74. Keeler, R.F., D. Brown, D. Douglas, G.F. Stallknecht
    and S. Young. 1976. Teratogenicity of Kennebec potato
    sprouts and solasodine in hamsters. Bull. Env.
    Contamin. and Toxic. 15:522.

75. Kadkade, P.G. and T.R. Madrid. 1977. Glycoalkaloids in
    tissue culture of Solanum acculeatissimum.
    Naturwissenschaften 64:147.

76. Sanford, L.L. and S.L. Sinden. 1972. Inheritance of
    potato glycoalkaloids. Amer. Pot. J. 49:209.

77. McCollum, G.D. and S.L. Sinden. 1979. Inheritance study
    of tuber glycoalkaloids in a wild potato, Solanum
    chacoense Bitter. Amer. Pot. J. 56:95.

Chapter Five

CORN KERNEL MODIFICATION*

EVELYN J. WEBER

U.S. Department of Agriculture
University of Illinois
Urbana, Illinois 61801

CORN PRODUCTION AND UTILIZATION

Corn is an excellent example of applied phytochemistry. Many genetic modifications of corn seeds have been studied, and a number of these modifications have had or may have economic value. The importance of corn to the economy of the United States can be illustrated by reviewing recent statistics.

Corn is the largest single grain crop produced in the United States. In 1978, production exceeded 7 billion bushels for the first time (Table 1). Corn production exceeded the combined totals of wheat, oats, soybeans, barley and rye. Soybeans and wheat each yielded 1.8 billion bushels in 1978, only one-fourth of the corn production.[1] The United States produced nearly one-half of

* Mention of a trademark name, proprietary product, or specific equipment does not constitute a guarantee or warranty by the U.S. Department of Agriculture and does not imply its approval to the exclusion of other products that may also be suitable.

Table 1.   U.S. Crop Production

Crop	1958	1978
	Billions of bushels	
Soybean	0.5	1.8
Wheat	1.1	1.8
Corn	3.8	7.1

Source: Parrott[1]

Table 2.   Corn: Production and Exports of Major
Exporting Countries in 1978

Country	Production	Exports
	Millions of bushels	
Brazil	747.8	11.8
France	370.0	90.5
Argentina	350.3	224.4
South Africa	299.1	70.8
Thailand	118.1	78.7
	1,885.3	476.2
United States	7,080.5	1,949.6
World Total	14,315.2	2,530.8

Source: USDA Foreign Agricultural Service
Feed and Grain Division

the corn grown in the world in 1978 (Table 2). Two states,
Iowa and Illinois, produced 38% of the U.S. corn crop or
nearly one and one-half times the corn produced by the five
leading foreign producers.

The value of agricultural exports to the U.S. is well
known. At present, these agricultural products account for
one-fourth of the nation's total export earnings. In 1978,
soybean exports were 775 million bushels; wheat exports 1.2
billion bushels; and corn exports 2 billion bushels.[1] The
United States' share of the world corn export market was
77% (Table 2). Our nearest competitor was Argentina, but
its share of the world export market was only 9%. During
the past 6 years, U.S. farm exports have tripled. Former
Secretary of Agriculture Clifford M. Hardin[2] has predicted
that our agricultural exports will continue to grow by
perhaps as much as 6 or 7% per year. Continued population
growth and rising affluence within many foreign countries
were the reasons given for the increasing demand for agricultural products.

In 1978 about one-third of U.S. production was exported
and two-thirds retained for domestic use. The yearly rate
of exports has been increasing faster than domestic use.
We need to continue developing new export markets, because
corn production is increasing faster than domestic use. In
the past 20 years (1958 to 1978) the production of corn has
nearly doubled (Table 1). In the Corn Belt, the annual
increase in yield has been about 2 1/2 bushels per acre per
year (1.6 q/ha/yr).[3] In 1931, before the use of hybrids
became widespread, the national average yield was 24.5 bu/a
(15.4 q/ha),[3] while in 1978 the average was 101 bu/a (63.3
q/ha). Some farmers participating in recent corn yield
contests have reported yields exceeding 300 bu/a (188
q/ha). The potential for continuing increases in average
yields certainly exists. The major factors contributing to
the higher yields have been the use of improved hybrids,
increased nitrogen fertilization, denser plant populations,
and effective weed and pest control.[3]

In considering modification of corn, we should think
carefully about the present major uses of corn. By far,
the major domestic use of corn is for livestock feed; 87%
was used for feed in the United States last year (Table 3).
Domestic feed usage is increasing slowly at present

Table 3.   Corn: Domestic Use

Year Beginning Oct. 1	Food	Alcoholic Beverages	Seed	Animal Feed
1974/75	366.9	65.7	18.8	3,225.6
1975/76	398.8	71.1	20.2	3,591.6
1976/77	419.4	73.9	19.8	3,586.6
1977/78	462.8	70.4	18.0	3,709.0
1978/79	488.8	68.0	18.0	4,000.0

Source: USDA Statistical Reporting Service

Table 4.   General Composition
of the Dry Dent-Corn Kernel

Component	%
Carbohydrates	80.0
Protein	10.0
Oil	4.5
Fiber	3.5
Minerals	2.0

Source: Jugenheimer[3]

because, although hog and poultry numbers are up, cattle numbers are still down.

The corn refining industry uses about 9% of the domestic corn supply. This market has doubled in the last 10 years from 207 million bushels in 1968 to about 400 million bushels in 1978.[4] Much of the recent growth in the corn refining industry has been due to the rapidly rising consumer demand for corn sweeteners. In 1978, sweeteners from corn (corn syrup, dextrose, and high-fructose corn syrup) captured over 25% of the market for nutritive sweeteners.[5] A decade ago, corn sweeteners only accounted for 15% of the market. Because corn sweeteners give the same perception of sweetness as sucrose but with fewer calories, their use in the manufacture of low-calorie foods is expected to increase dramatically.

This review, however, is not concerned with the thousands of food and industrial uses of corn but rather with the carbohydrates, protein, oil and other constituents that the corn plants produce biosynthetically. Corn is especially rich in natural genetic variability. Many special nutritional or industrial types of corn have been bred.

The general composition of the dry dent-corn kernel is shown in Table 4. The percentages of these components vary according to genotype and environmental conditions, but the values listed are typical for current commercial hybrids. Carbohydrates make up most of the corn kernel. Protein and oil levels are lower in corn than in soybean which has about 40% protein and 22% oil, but the average yield of corn is about three times that of soybean. Genetic modifications of corn which affect the quality and quantity of carbohydrate, protein, and oil already exist. Carbohydrate and protein mutants will be reviewed briefly; more information can be found in several recent books about corn.[3,6,7,8] Corn oil biosynthesis will be covered in more detail, as this is my research area.

MODIFICATION OF CORN KERNEL CONSTITUENTS

Carbohydrates

      In a cross section of the face of a corn kernel (Figure
1), we can identify parts of the kernel. The pericarp or
hull is the thin outer covering of the kernel. It consists
largely of carbohydrates, especially fiber or cellulose.
The tip cap originally attached the kernel to the cob.
Occasionally during shelling the tip cap is broken off the
kernel, but usually it remains on the kernel and serves as
a protective covering for the end of the germ. The hilar
layer or, as it is more commonly called, the black layer is
an impervious layer that forms at maturity and separates
the tissues of the kernel from the tip cap and cob. The
appearance of this black layer is often used as a guide to
determine physiological maturity, and it is considered a
more accurate indicator of maximum growth than moisture content or days after silking. The germ consists of the
scutellum and the embryo-axis. The scutellum is a feeding
organ for the embryo because of its rich supply of enzymes
and oil. The endosperm surrounds the germ on the sides and
in the back. Two types of endosperm tissue are recogniz-
able. The floury endosperm around the germ is soft, mealy
and relatively opaque. The horny endosperm is hard and
translucent and contains a higher concentration of protein
than is found in the floury endosperm. In yellow corn
varieties, the horny endosperm has a much deeper color than
the floury. The proportion of horny to floury endosperm
depends upon the type and variety of the corn.

      In popcorn, the horny endosperm forms a thick shell
around a small central core of floury endosperm. Popping
ability appears to be conditioned by the proportion of the
horny endosperm where the starch grains are embedded in a
tough, elastic colloidal material that resists the steam
pressure generated within the grains upon heating until
this pressure reaches explosive force. About 14% moisture
is optimum for popping. Only 0.1% of the total corn
acreage in the United States is devoted to popcorn.[3]
Unfortunately, yield and popping expansion appear to be
negatively correlated.[9] The popping-expansion trait is
polygenic and can be improved by selection. Weaver and
Thompson[10] reported that a 15-generation modified mass-
selection scheme increased popping expansion from 22.2 to
35.8 volumes in a White Hulless population. The presence

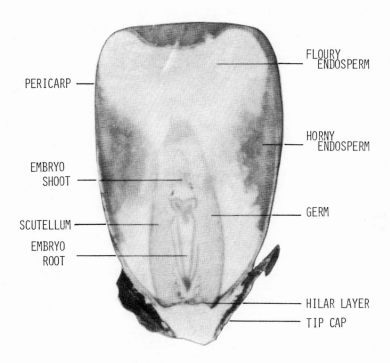

Figure 1.   Cross section of dent corn kernel.

of thick hulls in popcorn is objectionable to consumers.
Inheritance of pericarp thickness appears relatively
simple, with a single major dominant gene for thin pericarp
and a modifier complex that induces the production of thick
pericarp.[11] The biochemical bases for superior popping
ability and thinner pericarps in popcorn have not been
investigated.

The starch granule of standard dent corn usually contains a mixture of about 73% amylopectin and 27% amylose[12]
(Table 5). Amylopectin consists of chains of α-$\underline{D}$-(1-4)-
and α-$\underline{D}$-(1-6)-glucosidic linkages that form a branched
molecule of high molecular weight (about 1 x 10[7]). Amylose
is a linear chain polysaccharide with α-$\underline{D}$-(1-4)-linked glucose units. Bear[13] reported a single recessive gene that
doubled the amylose content of normal corn. The symbol $\underline{ae}$
for "amylose extender" was proposed for this gene.[14] A
consistent problem in high-amylose corn has been a reduction in the total amount of starch produced. Kernels homozygous for $\underline{ae}$ have a slightly shrunken appearance in many
genetic backgrounds. The shrunken condition is the result
of a partially collapsed endosperm which may, in turn, be
due to a reduction in amylopectin synthesis without a com-
pensating increase in amylose. Zuber $\underline{et}$ $\underline{al}$.[15] found that
amylose percentage was negatively correlated with endosperm
weight and starch percentage and positively correlated with
pericarp weight, germ weight, endosperm protein and
endosperm oil. Identification of the high-amylose mutants
has also been a problem. Sometimes a "translucent"
appearance of the endosperm or change in kernel color has
been noted, but these indicators are not as reliable as the
more time-consuming chemical analysis of the starch.[16]
Selection pressure for maximum amylose content,[17] and high
test weight,[18] has produced high-amylose hybrids that
approach the performance of normal Corn Belt hybrids.
These "amylomaizes" with 50 to 80% amylose are being produced and sold commercially on a small scale.[3] The special
physical and chemical properties of the high-amylose starch
are used for production of fibers, films and plastics.

A Presbyterian missionary in China sent the first
sample of waxy corn to the United States in 1908. The gene
was named waxy ($\underline{wx}$) by Collins,[19] because of the waxy
appearance of the endosperm. The starch produced by waxy
mutants is almost exclusively amylopectin, the branched
polysaccharide. Ordinary cornstarch stains blue with a

Table 5.   Amylose Content of Mature Kernels
           of 11 Genotypes of Corn

Gene Combination	Amylose in Starch (%)
Normal	27
ae	61
su1	29
su2	40
wx	0
ae su1	60
ae su2	54
ae wx	15
su1 su2	55
su1 wx	0
su2 wx	0

Source: Kramer, Pfahler & Whistler[12]

potassium iodide-iodine solution, whereas waxy cornstarch
stains a reddish brown. This staining reaction has been of
great aid to breeders, because the waxy gene also is
expressed in the pollen and can be detected by staining.
The introduction of the wx allele into standard, high-
yielding hybrids was initiated by G. F. Sprague and his
colleagues,[20] at Iowa State University. Production of waxy
corn has increased steadily. About 20 to 30 million
bushels a year are grown under contract to corn millers.[21]
Waxy corn makes up 8 to 10% of the corn processed by the
wet-milling industry. Amylopectin cornstarch has been used
in instant pudding mixes, glues and other products previously requiring tapioca (cassava) starch.

The wx gene completely suppresses amylose accumulation
in combinations with su1 or su2 genes but not with ae
(Table 5). These mutations that affect starch synthesis in
corn endosperm have been known for some time, but only in a
few instances has a biochemical lesion been identified with
a mutant. Nelson and Rines[22] found that starch granules
from wx endosperms were deficient in nucleoside diphosphate
glucose-starch glucosyl transferase. This enzyme catalyzes
the transfer of glucose to an α-D-(1-4) linkage on the
nonreducing end of the polysaccharide acceptor. This activity suggests that the dominant Wx allele may be involved
in the synthesis of amylose, but although both Wx dosage
and enzyme activity are doubled from diploid to tetraploid
maize, the amylose content remains the same.[23] Our
understanding of the control of starch synthesis in corn
endosperm is still very limited.

Relatively low amounts of two other enzymes have been
discovered in starch-deficient mutants. Shrunken-2 (sh2)
and brittle-2 (bt2) mutants have low (5-10%) adenosine
diphosphate glucose pyrophosphorylase activity (G-1-P + ATP $\rightleftharpoons$
ADPG + PP).[24,25] The residual activity may be due to a
pyrophosphorylase not coded by either of these loci,[26] or
null mutations may be lethal. The other enzymatic lesion
associated with a mutation has been reported recently by
Chourey and Nelson.[27] Shrunken-1 (sh1) mutants have only
10% of the normal sucrose synthetase activity. Biochemists
have not extensively used plant mutants to study metabolic
pathways. It may be possible to gain valuable information
about starch synthesis and other biochemical processes by
studying specific combinations of various mutations.

Ordinary sweet corn differs from field corn in only one recessive gene, sugary-1 (su1). In su1 kernels, water-soluble carbohydrate (phytoglycogen) accumulates at the expense of starch (Table 6). Laughnan introduced the sh2 gene into standard sweet-corn inbred lines.[28] The sh2 corn has less starch and higher levels of sugars than normal or sugary corn. The sh2 gene in combination with the su1 gene drastically reduces formation of starch and water-soluble polysaccharides and increases synthesis of reducing sugars and sucrose.

A single-cross sh2 hybrid, 'Illini X-tra Sweet' corn, is twice as sweet as normal sweet corn at harvest The conversion of sugar to starch after harvest is also slowed: if held for 48 hr at room temperature, the 'Illini X-tra Sweet' hybrid corn is four times as sweet as normal sweet corn.[3] One problem with the Illinois "supersweet" corns has been low germination of the seed.[29] In recent studies[30-32] various combinations of the endosperm-altering genes ae, wx, dull, (du), su1, su2, and sh2 have been incorporated into sweet-corn lines in a search for the improved germination, moisture, and sugar characteristics desired by the sweet-corn processing industry.

A newly developed inbred line, Illinois 677a, has a sucrose content equivalent to that of the su1 sh2 hybrids and a water-soluble polysaccharide level resembling that of the su1 corns.[33] Illinois 677a comes from a three-way cross: (Bolivia 1035 x Illinois 44b) x Illinois 442a. The Bolivia 1035 is an interlocking Coroico flour corn; Illinois 44b and Illinois 442a are su1 inbreds. The mechanism for the enhanced sucrose accumulation in Illinois 677a is not known, but it appears to be different from that of sh2. Illinois 677a does not have the drastically reduced levels of adenosine diphosphate glucose pyrophosphorylase that are observed in sh2 mutants.

Genes that improve protein quality are also being combined with the genes affecting carbohydrates in attempts to enhance the nutritional quality of sweet corn.

Protein

Commercial hybrid corn has only about 10% protein. Soybeans have 40% protein. Yet corn with its high yields and large acreage produced 92% as much total protein as

Table 6.   The Carbohydrate Composition of Kernels of
           Normal, Sugary (su1), and Shrunken (sh2) Corn

Genotype	Mean dry wt. per kernel mg	Reducing sugars %	Sucrose %	Water soluble polysac- charides %	Starch %	Total carbo- hydrates %
Normal	185	0.3	1.4	1.3	65.0	68.0
su1	166	1.9	2.7	35.8	30.0	70.3
sh2	139	2.7	16.1	1.6	24.8	45.2
su1 sh2	113	4.0	28.0	1.8	7.7	41.5

Source: Laughnan[28]

soybeans in the United States in 1978. The quantity and
quality of corn proteins are very important.

In 1896 at the University of Illinois, C. G. Hopkins[34]
started a series of experiments to determine whether the
protein and oil content of corn could be changed by
selection. A random sample of ears was drawn from a local
corn variety 'Burr's White'. Each ear was analyzed for
protein and oil content. Ears with the highest percentages
of protein and oil were selected to initiate the Illinois
High Protein (IHP) and Illinois High Oil (IHO) strains, and
ears with the lowest percentages, the Illinois Low Protein
(ILP) and Illinois Low Oil (ILO) strains. These selections
have been continued for more than 76 generations and are
considered classical examples of the effectiveness of
selection for chemical composition in corn kernels. The
original 'Burr's White' corn had 10.9% protein. After 76
generations of selection, the average protein content was
about 27% for IHP and 4% for ILP[35] (Figure 2). Selection
is still being carried on. The most important finding from
this long-term experiment is that significant genetic
variability for protein still exists in these strains.
Because no concurrent selection for yield was made, IHP now
produces only one-third and ILP one-half as much grain as
commercial hybrids,[36] but the high-protein trait can be
transferred to new lines with better yields by various
breeding procedures.[3,37]

Recently research has been focused on improving the
quality of corn protein. The major storage protein of corn
endosperm is zein which is deficient in lysine and tryptophan, essential amino acids for nonruminants such as swine,
chickens and humans. In normal corn, zein makes up 40-60%
of the total endosperm protein. Although the opaque-2 (o2)
mutant was recognized many years ago, the high quality of
its protein was not discovered until 1964, when Mertz,
Bates and Nelson reported that the o2 gene increased the
lysine content of the endosperm.[38] The net effect of the
o2 gene appears to be a partial repression of the synthesis
of the alcohol-soluble proteins, essentially zein, and an
oversynthesis of the other proteins and free amino acids[39]
(Table 7). Since the nonzein proteins have higher levels
of lysine and tryptophan than zein, these changes give o2
corn its improved nutritional quality.

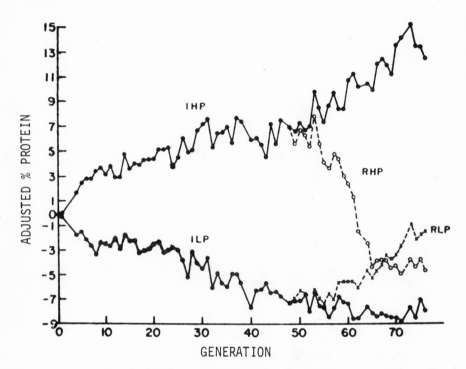

Figure 2. Mean adjusted percentage of protein for Illinois
High Protein (IHP), Illinois Low Protein (ILP),
Reverse High Protein (RHP), and Reverse Low
Protein (RLP) strains of corn plotted against
generations.[35] Varying environmental conditions
have caused large year to year fluctuations in
percent protein, so the mean protein values for
IHP, ILP, RHP, and RLP for a particular year
have been adjusted by substracting the mean percent protein of Illinois High Oil and Illinois
Low Oil for that year.

Table 7. Distribution of Soluble Protein Fractions
in Endosperms of Normal and opaque-2 Corn

| Extract | Protein fraction | Percentage of total protein wt | |
		Normal	o2
Water	Albumins[a]	3.7	14.7
5% NaCl	Globulins	1.7	4.4
70% EtOH	Prolamines[b]	54.2	25.4
0.2% NaOH	Glutelins	40.4	55.5
% lysine in total soluble protein		1.6	3.7

[a] - Includes free amino acids

[b] - Includes zein

Source: Jiminez[39]

Several other mutants have been identified that
decrease the ratio of zein to the other protein fractions.
In addition to increased lysine and tryptophan, the proteins of the floury 2 (fl2) endosperm contain about 60%
more methionine than those of normal corn.[40],[41] The fl2
gene is the only gene known in corn that causes a considerable increase in methionine. This is an important
discovery, because methionine is the limiting amino acid
for growth of man and monogastric animals on cereal diets
supplemented with legume protein such as soybean meal.
However, the fl2 gene generally produces lower lysine
levels than o2 and is considered inferior to o2 for
improving protein quality in corn.

The effects of the opaque-7,[42] opaque-6,[43] and
floury-3[43] genes are also similar to those of the other
high-lysine mutants in that the synthesis of zein is
suppressed and that of the free amino acids, albumin, and
insoluble proteins is markedly increased. The o6 and fl3
mutants do not appear to have practical value, however,
because o6/o6 plants die as seedlings and fl3 seeds are
very light in weight. All the high-lysine mutants have a
reduced endosperm weight. An increase in the percentage of
lysine may not represent an increase in yield of lysine if
the kernel size is greatly decreased.

Shrunken-4 (sh4) is a starch-deficient mutant that also
has a reduced zein content.[44] Mature kernels of sh4 contain only 10% of the normal content of zein per kernel; the
other protein mutants contain 25-50% of the normal zein
content. Whether this is a direct effect of the sh4 gene
on zein synthesis or a secondary consequence of the defec-
tive starch metabolism is not known. Burr and Nelson[45]
have shown that the sh4 endosperms contain about 1/8 as
much vitamin B6, pyridoxal phosphate, as do normal endosperms. Since pyridoxal phosphate is a cofactor for a
number of enzymes, the reduced quantity of the cofactor in
the sh4 mutant could account for the lower protein content.

Two other mutants, de*-91 and de*-92, have been
reported[46] in which the alcohol-soluble proteins and also
the albumins and globulins are greatly reduced.

Our present knowledge of protein synthesis in normal
corn endosperm and of the effects of the protein mutants on
the biosynthesis is limited. In normal corn, the alcohol-

soluble protein, zein, can be separated by SDS-polyacryla-
mide gel electrophoresis into two major components of about
19,000 and 22,800 molecular weight. The o2 mutant is deficient in the larger of these two protein components.[47]
Studies of the biochemical mechanisms that regulate zein
synthesis have indicated that part of the reduction in zein
synthetic capacity in the o2 mutant is due to a reduced
level of membrane-bound polyribosomal material.[48] The
mutant had only about one-half the amount of ribosomal RNA
found in the membrane-bound polysome fraction of normal
corn.

    When the albumins and globulins from o2 and normal
inbreds were separated by electrophoresis and compared, the
opaque inbreds did not show any characteristic changes.[49]
The o2 gene did not appear to affect any of the albumins
and globulins directly; rather, each inbred responded indi-
vidually by increasing or decreasing the relative proportions of particular proteins.

    Most of the current breeding programs for development
of better quality protein in corn involve the o2 gene.
Problems resulting from the introduction of the o2 gene
into hybrids have included lower yields and soft endosperms
that are more susceptible to ear rots, grain insects, and
damage during harvesting and storage.[50,51] Lambert et
al.[50] found that o2 hybrids averaged 8% lower in yield and
89% higher in cracked kernels than their normal dent
counterparts. Two hybrids in a group of 10, however, gave
yields that were not significantly different from their
normal counterparts. Dry-matter accumulation is suppressed
earlier in kernels of o2 than in their normal counterparts;
however, the length of time of suppression is a function of
genetic background.[52,53] Modifier genes have also been
detected that produce harder kernels in o2 inbreds with
little or no reduction in protein quality.[54,55] A harder
kernel would presumably reduce harvesting and storage
damage. Problems with poor germination and seedling vigor
of the o2 genotypes have slowed progress in the development
of improved o2 strains, but seedling emergence can be
improved by selection without harming protein quality.[56]
All of the agronomic deficiencies associated with o2 corn
appear to be amenable to improvement by breeding. It may
be possible to breed corn with high-quality protein that
can compete favorably with ordinary hybrids.

Lipids

    Selection for higher and lower oil content in corn was
included along with selection for protein in the classical
plant breeding experiment started by Hopkins at the University of Illinois in 1896.[34] The original 'Burr's White'
ears averaged 4.7% oil. After 76 generations of selection
of ears with highest oil content, the oil content in the
Illinois High Oil (IHO) strain has increased to 19%[35]
(Figure 3). The Illinois Low Oil (ILO) has a mean of 0.3%
oil. Progress in ILO may be near an end, because many of
these kernels do not have germs.

    The reverse selections, Reverse Low Oil (RLO) and
Reverse High Oil (RHO), were initiated after 58 generations
to determine whether genetic variation remained. The procedure was the same as that in the forward-selection
experiment, except that selection was for high oil in the
low oil strain and low oil in the high oil strain.
Switchback High Oil (SHO) was started from RHO after seven
generations of reverse selection. The progress realized in
all of the strains indicated that the long-continued selection did not result in fixation of the genes governing high
or low oil content. Even in IHO, oil percentage should
increase further if selection is continued. At 19% oil,
IHO is within the range of oil in oilseeds such as soybeans, which have 17-22% oil. However, as with the IHP
strain, the yield of IHO has fallen drastically.

    The energy content of oil is 9 kcal/g; that of protein
or carbohydrate is only 4 kcal/g. If the higher oil strains
yielded as much grain per acre as the commercial hybrids,
the higher oil types would obviously produce more calories
per acre. Without selection for yield, IHO with 17% oil
after 70 generations of selection produced only 42 bu/a[57]
(Table 8). A common hybrid, Mo17 x N28, had 3.2% oil and
yielded 179 bu/a.[58] Adaptation of wide-line nuclear-
magnetic-resonance (NMR) spectroscopy to nondestructive
analysis of oil content in single corn kernels[59-61]   has
made selection for higher oil content more efficient. An
oil analysis can be carried out in 2 seconds.[58] Selection
for both oil content and yield has produced hybrids with
6.4 - 8.8% oil and grain yields equal to those of good commercial hybrids[58] (Table 8).

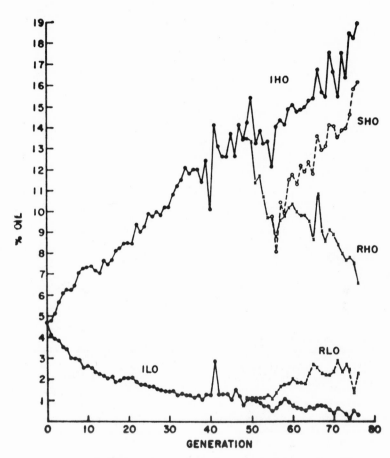

Figure 3. Mean percentage of oil for Illinois High Oil
           (IHO), Illinois Low Oil (ILO), Reverse High Oil
           (RHO), Reverse Low Oil (RLP), and Switch-back
           High Oil (SHO) strains of corn plotted against
           generations.[35]

Table 8. Performance of High Oil and
          Commercial Corn Hybrids

Entry	Yield bu/a	Oil %
Illinois High Oil	42	17.0
R802A x R806	166	8.8
R806 x H49HO	189	8.0
R806 x N28	173	6.4
Mo17 x N28	179	3.2
Commercial single cross	181	3.9
Highest-yielding commercial hybrid	187	4.0

Source: Dudley, Lambert & Alexander[57]
         Creech & Alexander[58]

Because corn is largely fed to livestock, an increase in the energy value of corn would be expected to be advantageous. Animal feeding experiments with higher-oil corns, however, have had variable results. One feeding experiment with growing-finishing pigs[62] showed that 5-6% less high-oil (7% oil) corn than normal corn was required per pound of gain. The protein supplement was also reduced, because the higher-oil corns have increased germ size and tend to have slightly higher levels of protein. The quality of germ protein also is better than that of the zein from endosperm. Another swine feeding experiment[63] showed no advantage of 5.9% oil over 4% oil corn. Pigs do not seem to utilize the full caloric value of corn as indicated by bomb calorimeter tests. The form in which the corn is fed, such as meal or pellets, affects the availability of the calories. Further experiments are needed to determine what level of oil in corn would be optimum in livestock feed.

Corn millers should find higher-oil corn profitable, because corn oil is in strong demand and has the highest value per unit weight of any fraction of the corn kernel. The development of low-cost, rapid infrared grain analyzers[64] may solve the problem of field identification of higher-oil and higher-protein corns. If the analyzer was used at a grain elevator, the farmer could be paid a premium price for corn with higher than normal oil or protein.

The major selling point for corn oil has been its high level of polyunsaturated fatty acid. Corn oil from commercial hybrids contains over 60% diunsaturated linoleic acid (Table 9). Among the common vegetable oils, only sunflower oil has a higher percentage of linoleic acid: soybean, cottonseed and peanut oils have lower levels than corn. Cottonseed and palm oils and the animal fat, lard, have more saturated fatty acids. Another advantage of corn oil over soybean oil is that corn oil contains a lower level of the triunsaturated linolenic acid, which oxidizes readily to produce off-flavors.[68]

In the past 15 years, the linoleic acid level of corn oil has increased 3-5%.[69] This change was accidental and could just as well have been in the opposite direction. New hybrids were selected because they had superior agronomic performance and not because they had desirable oil

Table 9. Fatty Acid Composition of Commercial
Oils and Fats before Hydrogenation

Oil	Fatty acid composition (mol %)				
	16:0[a]	18:0	18:1	18:2	18:3
Sunflower	6.3	3.8	25.7	64.2	--[b]
Corn	12.0	2.0	23.8	61.4	0.8
Soybean	11.3	3.1	22.0	54.1	8.3
Cottonseed	26.6	2.7	16.6	51.6	--
Peanut	11.9	2.3	50.4	30.6	--
Lard	28.8	13.0	42.4	10.3	0.5
Palm	49.0	3.6	35.9	9.6	--

[a] Fatty acids are identified according to number of carbon
atoms and number of double bonds - palmitic (16:0),
stearic (18:0), oleic (18:1), linoleic (18:2), and
linolenic (18:3).

[b] No values reported.

Sources: Reiners and Gooding[65]
Weiss[66]
Morrison and Robertson[67]

quality. Commercial breeders seldom monitor the fatty acid composition of corn oil.

Recently questions have arisen about the wisdom of recommending a continuing increase in the consumption of polyunsaturates for the general public.[70,71] This dietary change has not been conclusively shown to prevent atherosclerosis. A possible risk factor associated with consumption of polyunsaturated oils is that their oxidation products, lipid peroxides, have been implicated as carcinogenic agents.

Another current controversy involving polyunsaturated vegetable oils concerns the physiological effects of dietary trans fatty acids.[72,73] During hydrogenation of vegetable oils, the natural cis form of the double bonds of the fatty acids may be converted to the trans form. In the more solid products, the vegetable oils are more highly hydrogenated, and the content of trans acids is increased. For example, in salad oils, the content of trans acids is only 0-15%, but in stick margarines, 25-35% of the fatty acids are in the trans form.[72] The use of a more saturated oil would reduce the hydrogenation required to form margarines and solid shortenings.

Whatever the outcome of all of these debates, corn has great genetic diversity in fatty acid composition that can be used to modify the polyunsaturation of its oil. The fatty acid compositions of six strains are shown in Table 10. Linoleic acid ranges from 28.9-69.6%. Linoleic and oleic acids usually comprise 80-90% of the total fatty acids in corn oil and are negatively correlated. Thus the low linoleic acid (28.9%) strain has high oleic acid (57.2%). Jellum[74] found even greater diversity in the fatty acid compositions of corn from countries other than the United States; the extremes for linoleic acid were 19-71%. Selection for a desired fatty-acid composition in corn oil should be possible because the levels of oleic and linoleic acid are highly heritable.[75-78] Genotype has a much greater influence on fatty acid composition than do environmental factors such as temperature, planting date or fertility.[79,80]

The fatty acids of corn oil are bound in triglycerides. Three fatty acids are esterified to a glycerol molecule. Over the years, there have been many theories as to how the

Table 10. Fatty Acid Composition of Oil
from Six Strains of Corn

Strain	Fatty acid composition (mol %)				
	16:0	18:0	18:1	18:2	18:3
C105-L	10.3	2.6	57.2	28.9	1.0
IHO	12.0	2.0	37.0	48.2	0.8
B37	14.8	1.2	23.1	59.6	1.4
A632	11.0	1.1	20.5	66.1	1.3
Mo17	11.0	1.6	19.4	67.2	0.8
NY16	6.5	1.3	21.4	69.6	1.2

Source: Weber - Unpublished data

fatty acids were distributed among these three positions.
The simplest was the random distribution theory in which
the fatty acid composition at each of the three positions
would be identical. No experimental data were available
for many years because there was no way to determine the
fatty acid composition at each position. Brockerhoff[81] and
Lands et al.[82] developed procedures in which stereospecific
enzymes were used to analyze the fatty acids at each
position. When animal triglycerides were analyzed,[82-84]
the fatty acids were found not to be distributed at random,
i.e., the fatty acid composition at each position was
different. Corn oil, soybean oil and several other vegetable oils were also stereochemically analyzed.[85] In the
vegetable oils, the fatty acid composition at the middle or
2-position was distinctly different from that at either
outer position (1- or 3-), but the fatty acid compositions
at 1 and 3 were very similar. There was still some
question as to whether the distribution of fatty acids was
random between the 1- and 3-positions of triglycerides from
plants.

We noticed that all of these analyses had been done on
commercial vegetable oils (Mazola corn oil, for example).
When we analyzed the triglycerides from corn inbreds,[86] we
found significant differences in fatty acid distribution at
all three positions (Table 11). The fatty acid distribution was clearly nonrandom at the 1-, 2-, and 3-positions.
The saturated fatty acids were predominantly esterified at
the 1-position. In the 2-position, more than 98% of the
fatty acids were unsaturated. The percentage of saturated
fatty acids was higher in position 1 than in position 3,
and the difference in position 3 was made up by oleic acid
or linoleic acid or both. The most interesting finding of
these studies was that each inbred had its own characteristic fatty acid pattern. For example, in H51 and C103,
the percentage of oleic acid (18:1) was higher in 3 than in
1, but in NY16, it was higher in 1. In H51, the level of
linoleic acid was higher in 1 than in 3, whereas in the
other two inbreds, it was higher in 3 than in 1. Crosses
of inbreds indicated some heritability of fatty acid placement within the triglyceride molecule.[87] Table 11 shows
the fatty acid patterns of C103 and NY16 and their cross,
C103 x NY16. The percentages of fatty acids at each position of the triglycerides of the cross were intermediate
between those of the parents.

Table 11.   Stereospecific Analyses of Triglycerides from
            Three Corn Inbreds and a Cross of Two of the Inbreds

		Fatty acid distribution (mol %)				
Strain	Position	16:0	18:0	18:1	18:2	18:3
H51	1	26.0	3.4	30.8	38.8	1.0
	2	1.5	0.1	26.8	70.6	1.0
	3	25.4	2.0	36.1	34.9	1.6
C103	1	22.4	4.2	41.2	31.7	0.5
	2	1.0	0.3	40.4	57.5	0.8
	3	14.2	2.1	47.5	35.6	0.6
C103 x NY16	1	21.3	3.2	30.4	44.3	0.8
	2	0.6	0.2	27.6	70.8	0.7
	3	9.0	1.5	38.4	50.1	1.0
NY16	1	15.6	3.9	21.4	57.8	1.3
	2	0.7	0.2	21.6	76.6	0.8
	3	7.0	1.6	19.2	70.6	1.6

Source: Weber, de la Roche and Alexander[86]

The particular type of fatty acid at each position of the triglyceride is important for two very practical reasons. First, the fatty acids in the outer positions (1- and 3-) of the triglyceride are more susceptible to oxidation[88-90] and hydrogenation[91] than the fatty acid in the 2-position. Second, animals incorporate into their lipids largely the fatty acids at the middle position of dietary triglycerides.[92] During digestion of a triglyceride, pancreatic lipase cleaves the fatty acids from the outer positions, and the intestinal wall absorbs the resulting 2-monoglyceride. The free fatty acids are metabolized or reesterified. The fatty acid in the monoglyceride remains bound and is resynthesized into triglycerides or other lipids.

In summary, we have found that oil in corn can be modified by breeding in at least three ways: (a) quantity of the oil; (b) fatty acid composition of the oil; and (c) fatty acid placement within the triglyceride.

At present, we are looking at the fatty acid compositions not only of the triglycerides (oil) but also of the phospholipids.[69] We have found that each lipid class has a characteristic fatty acid pattern (Table 12). The triglycerides have low percentages of saturated palmitic acid and high percentages of linoleic acid. The phosphatidyl-cholines have the highest levels of oleic acid. Among the phospholipids, the phosphatidylethanolamines have the highest percentages of linoleic acid. Both the phosphatidylinositols and phosphatidylglycerols have high percentages of saturated fatty acids, but the phosphatidylglycerols tended to have more oleic and less linoleic than the phosphatidylinositols.

The genotype of the inbred superimposes variations in fatty acid composition within the characteristic lipid-class patterns. When the inbreds are ranked in the order shown in Table 12, increasing levels of linoleic acid generally are noted in each class of lipid. The range is much larger for the triglycerides, from 42.2% for H21 to 69.6% for NY16, but the differences also are apparent in the other lipid classes.

Preliminary results indicate that the fatty acid compositions of phospholipids are genetically controlled.

Table 12. Fatty Acid Composition of Classes of Lipids
           from Kernels of Four Corn Inbreds

Lipid Class	Inbred	Fatty acid composition (mol %)				
		16:0	18:0	18:1	18:2	18:3
Triglyceride	H21	16.5	2.9	37.4	42.2	1.0
	IHO	12.9	2.1	35.4	48.8	0.8
	K6	11.3	1.1	22.1	64.1	1.3
	NY16	7.4	1.6	20.1	69.6	1.3
Phosphatidyl-	H21	20.0	1.5	42.1	36.0	0.5
choline	IHO	19.0	1.5	37.5	41.4	0.6
	K6	23.9	1.8	28.3	45.2	0.8
	NY16	18.6	2.2	25.8	52.3	1.1
Phosphatidyl-	H21	42.1	2.6	18.7	35.6	1.0
inositol	IHO	39.5	2.2	20.6	37.4	0.3
	K6	40.2	1.9	12.3	44.1	1.5
	NY16	38.3	2.2	12.8	45.8	1.0
Phosphatidyl-	H21	20.8	0.8	23.5	54.2	0.7
ethanolamine	IHO	25.1	1.0	18.9	54.3	0.7
	K6	24.0	1.0	15.3	59.0	0.7
	NY16	20.6	1.9	18.1	58.3	1.0
Phosphatidyl-	H21	36.3	2.0	19.7	40.3	1.7
glycerol	IHO	36.7	2.7	23.6	36.6	0.5
	K6	35.5	2.5	15.6	42.6	3.8
	NY16	34.2	3.6	16.6	44.3	1.4

Source: Weber[69]

Reciprocal crosses were made of the inbreds, C103D and B73, which differ widely in linoleic acid content (Table 13). The lipids were isolated only from the germ, which has equal inheritance from both parents, and lipids from the triploid endosperm were excluded. The triglycerides and the phospholipids of the crosses showed intermediate linoleate values possibly with maternal effects. In the summer of 1979 we hope to produce the F_2 and reciprocal backcross generations of these and other inbreds so that we may further investigate the inheritance of fatty acids in phospholipids. The effect of genotype on phospholipids may have important consequences. Phospholipids are integral components of membranes, which determine the ability of the seed to resist desiccation,[93] cold,[94,95] and insects.[96] Changes in the fatty acid composition of the phospholipids may alter these properties.

Other Constituents

Genetic variations may exist in many components of corn in addition to the carbohydrates, proteins and lipids. We are currently analyzing the relative concentrations of the tocopherol isomers in various inbreds (Table 14). A wide range of α- and γ-tocopherols exist among these inbreds. The γ-isomer has usually been considered to predominate in corn, but B14 and B37 have higher levels of α than of γ. A recent paper[97] suggested that γ-tocopherol may be a better antioxidant than α-tocopherol for linoleic acid. We are concerned about the vitamin E isomers, because in breeding high-oil corn, it may be necessary to consider the level of tocopherols needed to protect the polyunsaturates during storage of the grain.

In conclusion, I would like to make two points. The first is that our knowledge of starch, protein and oil biosynthesis in corn and other plants is relatively limited, although these are areas of considerable practical importance. The second is that we have a large number of mutants that could serve as experimental probes to facilitate coordinated biochemical and genetic studies. Genetic variations have been rigorously identified at perhaps 350 loci in corn and loosely defined at perhaps 1000 or more loci.[98] Many of these involve visible traits. We undoubtedly have a tremendous wealth of biochemical variations that have not even been discovered. Many more modifications of corn may be possible in the future.

Table 13. Fatty Acid Compositions of Lipids
from Corn Inbreds and Crosses

Lipid	Inbred or Cross	Fatty acid composition (mol %)				
		16:0	18:0	18:1	18:2	18:3
Triglyceride	C103D	13.3	2.0	43.4	40.3	1.1
	C x B	12.6	2.1	38.6	44.7	1.0
	B x C	12.2	2.2	33.7	50.8	1.0
	B73	11.4	1.9	29.6	55.9	1.3
Phosphatidyl-choline	C103D	20.8	2.3	49.5	26.6	0.7
	C x B	19.1	1.8	44.1	34.2	0.8
	B x C	20.1	1.8	37.6	39.6	0.9
	B73	20.9	1.6	30.5	45.9	1.1
Phosphatidyl-inositol	C103D	38.9	4.0	29.8	26.8	0.5
	C x B	36.3	2.8	26.8	33.5	0.7
	B x C	37.5	2.5	20.9	38.4	0.7
	B73	33.7	2.4	19.8	42.9	1.2
Phosphatidyl-ethanolamine	C103D	23.0	3.9	38.4	34.2	0.5
	C x B	20.7	3.3	34.0	41.6	0.4
	B x C	20.6	2.2	26.7	49.8	0.7
	B73	21.5	2.7	21.9	52.8	1.0

Source: Weber - Unpublished data.

Table 14. Tocopherols in Corn Grain

Inbred	Tocopherols (µg/g)		Ratio
	α	γ	/α
B14	8.4	4.8	0.6
B37	33.0	19.5	0.6
A619	28.5	67.5	2.4
R802A	11.2	36.8	3.3
W64A	3.9	36.7	9.4

Source: Yen α Weber - Unpublished data.

REFERENCES

1. Parrott, R.B. 1979. Changing horizons of U.S. agriculture. Cereal Foods World 24:176-179.

2. Hardin, C.M. 1979. Research and agriculture. Cereal Foods World 24:175.

3. Jugenheimer, R.W. 1976. Corn Improvement, Seed Production and Uses. John Wiley & Sons, Inc. (New York, NY).

4. Curry, J. 1979. Corn agriculture, USA. In: Corn Annual. Corn Refiners Association, Inc. (Wash., DC), pp. 12-15.

5. Harrington, W.R. 1979. Introduction. In: Corn Annual. Corn Refiners Association, Inc. (Wash., DC), pp. 4-5.

6. Bauman, L.F., E.T. Mertz, A. Caballo and E.W. Sprague, (eds). 1975. High Quality Protein Maize. Dowden, Hutchinson and Ross, Inc. (Stroudsburg, PA).

7. Sprague, G.F., (ed.). 1977. Corn and Corn Improvement. American Society of Agronomy, Inc. (Madison, WI).

8. Walden, D.B., (ed.). 1978. Maize Breeding and Genetics. John Wiley & Sons. (New York, NY).

9. Alexander, D.E. and R.G. Creech. 1977. Breeding special industrial and nutritional types. In: Corn and Corn Improvement. (G.F. Sprague, ed.) American Society of Agronomy, Inc., Madison, WI. pp. 363-390.

10. Weaver, B.L. and A.E. Thompson. 1957. Fifteen generations of selection for improved popping expansion in White Hulless popcorn. Illinois Agricultural Experiment Station Bulletin 616.

11. Richardson, D.L. 1959. Pericarp thickness in popcorn. Agron. J. 52:77-80.

12. Kramer, H.H., P.L. Pfahler and R.L. Whistler. 1958. Gene interactions in maize affecting endosperm properties. Agron. J. 50:207-210.

13. Bear, R.P. 1958. The story of amylomaize hybrids. Chemurgic Digest 17:5.

14. Vineyard, M.L. and R.P. Bear. 1952. Amylose content. Maize Genetics Coop. News Letter 26:5.

15. Zuber, M.S., W.L. Deatherage, C.O. Grogan and M.M. MacMasters. 1960. Chemical composition of kernel fractions of corn samples varying in amylose content. Agron. J. 52:572-575.

16. Haunold, A. and M.F. Lindsey. 1964. Amylose analysis of single kernels and its implication for breeding of high-amylose corn. Crop Sci. 4:58-60.

17. Helm, J.L., V.L. Fergason and M.S. Zuber. 1967. Development of high amylose corn by the backcross method. Crop. Sci. 7:659-663.

18. Helm, J.L., A.V. Paez, P.J. Loesch, Jr. and M.S. Zuber. 1971. Test weight in high amylose corn. Crop Sci. 11:75-77.

19. Collins, G.N. 1909. A new type of Indian corn from China. U.S. Dept. of Agric. Bulletin 161.

20. Sprague, G.F. and M.T. Jenkins. 1948. The development of waxy corn for industrial use. Iowa State College J. Sci. 22:205-213.

21. Watson, S.A. 1977. Industrial utilization of corn. In: Corn and Corn Improvement. (G.F. Sprague, ed.). American Society of Agronomy, Inc. (Madison, WI), pp. 721-763.

22. Nelson, O.E. and H.W. Rines. 1962. The enzymatic deficiency in the waxy mutant of maize. Biochem. Biophys. Res. Commun. 9:297-300.

23. Nelson, O.E., Jr. 1978. Gene action and endosperm development in maize. In: Maize Breeding and Genetics. (D.B. Walden, ed.). John Wiley & Sons (New York, NY), pp. 389-403.

24. Tsai, C.-Y. and O.E. Nelson. 1966. Starch-deficient maize mutant lacking adenosine diphosphate glucose pyrophosphorylase activity. Science 151:341-343.

25. Dickinson, D.B. and J. Preiss. 1969. Presence of ADP-glucose pyrophosphorylase activity in shrunken-2 and brittle-2 mutants of maize endosperm. Plant Physiol. 44:1058-1062.

26. Hannah, L.C. and O.E. Nelson. 1975. Characterization of adenosine diphosphate glucose pyrophosphorylases from developing maize seeds. Plant Physiol. 55:297-302.

27. Chourey, P.S. and O.E. Nelson. 1976. The enzymatic deficiency conditioned by the shrunken-1 mutation in maize. Biochem. Genetics 14:1041-1055.

28. Laughnan, J.R. 1953. The effect of the sh2 factor on carbohydrate reserves in the mature endosperm of maize. Genetics 38:485-499.

29. Nass, H.G. and P.L. Crane. 1970. Effect of endosperm mutants on germination and early seedling growth rate in maize (Zea mays L.). Crop Sci. 10:139-140.

30. Rowe, D.E. and D.L. Garwood. 1978. Effects of four maize endosperm mutants on kernel vigor. Crop Sci. 18:709-712.

31. Soberalske, R.M. and R.H. Andrew. 1978. Gene effects on kernel moisture and sugars of near-isogenic lines of sweet corn. Crop. Sci. 18:743-746.

32. Andrew, R.H. and J.H. von Elbe. 1979. Processing potential for diallel hybrids of high-sugar corn. Crop Sci. 19:216-218.

33. Gonzales, J.S., A.M. Rhodes and D.B. Dickinson. 1976. Carbohydrate and enzymic characterization of a high sucrose sugary inbred line of sweet corn. Plant Physiol. 58:28-32.

34. Hopkins, C.G. 1899. Improvement in the chemical composition of the corn kernel. Illinois Agric. Exp. Station Bulletin 55:205-240.

35. Dudley, J.W. 1977. Seventy-six generations of selection for oil and protein percentage in maize. In: Proc. Int'l. Conference on Quantitative Genetics. (E. Pollak, O. Kempthorne and T.B. Bailey, Jr., eds), Iowa State University Press, (Ames, IA), pp. 459-473.

36. Dudley, J.W., R.J. Lambert and I.A. de la Roche. 1977. Genetic analysis of crosses among corn strains divergently selected for percent oil and protein. Crop Sci. 17:111-117.

37. Pollmer, W.G., D. Eberhard and D. Klein. 1978. Inheritance of protein and yield of grain and stover in maize. Crop. Sci. 18:757-759.

38. Mertz, E.T., L.S. Bates and O.E. Nelson. 1964. Mutant gene that changes protein composition and increases lysine content of maize endosperm. Science 145:279-280.

39. Jiminez, J.R. 1966. Protein fractionation studies of high lysine corn. In: Proc. of the High Lysine Corn Conference, Purdue University. (E.T. Mertz and O.E. Nelson, eds.), Corn Industries Research Foundation. (Wash. DC), pp. 74-79.

40. Nelson, O.E., E.T. Mertz and L.S. Bates. 1965. Second mutant gene affecting the amino acid pattern of maize endosperm proteins. Science 150:1469-1470.

41. Hansel, L.W., C.-Y. Tsai and O.E. Nelson. 1973. The effect of the floury-2 gene on the distribution of protein fractions and methionine in maize endosperm. Cereal Chem. 50:383-394.

42. McWhirter, K.S. 1971. A floury endosperm, high lysine locus on chromosome 10. Maize Genetics Coop. News Letter 45:184-185.

43. Ma, Y. and O.E. Nelson. 1975. Amino acid composition and storage proteins in two new high-lysine mutants in maize. Cereal Chem. 52:412-419.

44. Tsai, C.-Y. and A. Dalby. 1974. Comparison of the effect of shrunken-4, opaque-2, opaque-7, and floury-2 genes on the zein content of maize during endosperm development. Cereal Chem. 51:825-829.

45. Burr, B. and O.E. Nelson. 1973. The phosphorylases of developing maize seeds. Ann. N.Y. Acad. Sci. 210:129-138.

46. Foard, D., Y. Ma and O.E. Nelson. 1974. The mutations de*-91 and de*-92. Maize Genetics Coop. News Letter 48:169-172.

47. Lee, K.H., R.A. Jones, A. Dalby and C.-Y. Tsai. 1976. Genetic regulation of storage protein content in maize endosperm. Biochem. Genetics 14:641-650.

48. Jones, R.A., B.A. Larkins and C.-Y. Tsai. 1977. Storage protein synthesis in maize. II. Reduced synthesis of a major zein component by the opaque-2 mutant of maize. Plant Physiol. 59:525-529.

49. Wilson, C.M. 1978. Some biochemical indicators of genetic and developmental controls in endosperm. In: Maize Breeding and Genetics. (D.B. Walden, ed.), John Wiley & Sons, (New York, NY), pp. 405-419.

50. Lambert, R.J., D.E. Alexander and J.W. Dudley. 1969. Relative performance of normal and modified protein (opaque-2) maize hybrids. Crop Sci. 9:242-243.

51. Brown, W.L. 1975. Worldwide industry experience with opaque-2 maize. In: High Quality Protein Maize. (L.F. Bauman, E.T. Mertz, A. Caballo and E.W. Sprague, eds.) Dowden, Hutchinson and Ross, Inc. (Stroudsburg, PA), pp. 256-264.

52. Arnold, J.M., L.F. Bauman and D. Makonnen. 1977. Physical and chemical kernel characteristics of normal and opaque-2 endosperm maize hybrids. Crop Sci. 17:362-366.

53. Baenziger, P.S. and D.V. Glover. 1979. Dry matter accumulation in maize hybrids near isogenic for endosperm mutants conditioning protein quality. Crop Sci. 19:345-349.

54. Dalby, A. and C.-Y. Tsai. 1974. Zein accumulation in phenotypically modified lines of opaque-2 maize. Cereal Chem. 51:821-825.

55. Vasal, S.K. 1975. Use of genetic modifiers to obtain
    normal-type kernels with the opaque-2 gene. In: High-
    Quality Protein Maize. (L.F. Bauman, E.T. Mertz, A.
    Caballo and E.W. Sprague, eds.), Dowden, Hutchinson
    and Ross, Inc. (Stroudsburg, PA), pp. 197-216.

56. Loesch, P.J., Jr., W.J. Wiser and G.D. Booth. 1978.
    Emergence comparisons between opaque and normal
    segregates in two maize synthetics. Crop Sci.
    18:802-805.

57. Dudley, J.W., R.J. Lambert and D.E. Alexander. 1974.
    Seventy generations of selection for oil and protein
    concentration in the maize kernel. In: Seventy
    Generations of Selection for Oil and Protein in
    Maize. (J.W. Dudley, ed.) Crop Science Society of
    America, (Madison, WI), pp. 181-212.

58. Creech, R.G. and D.E. Alexander. 1978. Breeding for
    industrial and nutritional quality in maize. In:
    Maize Breeding and Genetics. (D.B. Walden, ed.),
    John Wiley & Sons (New York, NY), pp. 249-264.

59. Bauman, L.F., T.F. Conway and S.A. Watson. 1963.
    Heritability of variations in oil content of indivi-
    dual corn kernels. Science 139:498-499.

60. Alexander, D.E., L.S. Silvela, F.I. Collins and R.C.
    Rodgers. 1967. Analysis of oil content of maize by
    wide-line NMR. J. Amer. Oil Chem. Soc. 44:555-558.

61. Watson, S.A. and J.E. Freeman. 1975. Breeding corn for
    increased oil content. In: Proc. 30th Ann. Corn and
    Sorghum Res. Conf., Amer. Seed Trade Assoc. (Wash.,
    DC), pp. 251-275.

62. Nordstrum, J..W., B.R. Behrends, R.J. Meade and E.H.
    Thompson. 1972. Effects of feeding high oil corns to
    growing-finishing swine. J. Animal Sci. 25:357-361.

63. Lynch, P.B., D.H. Baker, B.G. Harmon and A.H. Jensen.
    1972. Feeding value for growing-finishing swine of
    corns of different oil contents. J. Animal Sci.
    35:1108.

64. Hymowitz, T., J.W. Dudley, F.I. Collins and C.M. Brown. 1974. Estimation of protein and oil concentration in corn, soybean and oat seed by near-infrared light reflectance. Crop Sci. 14:713-715.

65. Reiners, R.A. and C.M. Gooding. 1970. Corn oil. In: Corn: Culture, Processing, Products. (G.E. Inglett, ed.) Avi Publ. Co. (Westport, CT), pp. 241-261.

66. Weiss, T.J. 1970. Food Oils and Their Uses. Avi Publ. Co. (Westport, CT).

67. Morrison, W.H., III and J.A. Robertson. 1978. Hydrogenated sunflower oil: oxidative stability and polymer formation on heating. J. Amer. Oil Chem. Soc. 55:451-453.

68. Ho, C.-T., M.S. Smagula and S.S. Chang. 1978. The synthesis of 2-(1-pentenyl) furan and its relationship to the reversion flavor of soybean oil. J. Amer. Oil Chem. Soc. 55:233-237.

69. Weber, E.J. 1978. Corn lipids. Cereal Chem. 55:572-584.

70. Albrink, M.J. 1974. Serum lipids, diet and cardiovascular disease. Postgrad. Med. 55:87-92.

71. West, C.E. and T.G. Redgrave. 1974. Reservations on the use of polyunsaturated fats in human nutrition. Search 5:90-94.

72. Kummerow, F.A. 1975. Lipids in atherosclerosis. J. Food Sci. 40:12-17.

73. Kaunitz, H. 1976. Biological effects of trans fatty acids. Z. Ernährungswiss. 15:26-33.

74. Jellum, M.D. 1970. Plant introductions of maize as a source of oil with unusual fatty acid composition. J. Agr. Food Chem. 18:365-370.

75. Poneleit, C.G. and D.E. Alexander. 1965. Inheritance of linoleic and oleic acids in maize. Science 147:1585-1586.

76. Poneleit, C.G. and L.F. Bauman. 1970. Diallel analyses
    of fatty acids in corn (Zea mays L.) oil. Crop Sci.
    10:338-341.

77. de la Roche, I.A., D.E. Alexander and E.J. Weber. 1971.
    Inheritance of oleic and linoleic acids in Zea mays
    L. Crop Sci. 11:856-859.

78. Widstrom, N.W. and M.D. Jellum. 1975. Inheritance of
    kernel fatty acid composition among six maize
    inbreds. Crop Sci. 15:44-46.

79. Jellum, M.D. and J.E. Marion. 1966. Factors affecting
    oil content and oil composition of corn (Zea mays L.)
    grain. Crop Sci. 6:41-42.

80. Jahn-deesbach, W., R. Marquard and M. Heil. 1975.
    Investigations concerning fat quality in corn of
    German derivation with special consideration of lino-
    leic acid content. Z. Lebensm. Unters. Forsch.
    159:271-278.

81. Brockerhoff, H. 1966. A stereospecific analysis of
    triglycerides. J. Lipid Res. 6:10-15.

82. Lands, W.E.M., R.A. Pieringer, P.M. Slakey and A.
    Zschocke. 1966. A micro-method for stereospecific
    determination of triglyceride structure. Lipids
    1:444-448.

83. Brockerhoff, H., R.J. Hoyle and N. Wolmark. 1966.
    Positional distribution of fatty acids in triglycer-
    ides of animal depot fats. Biochim. Biophys. Acta
    116:67-72.

84. Brockerhoff, H., R.J. Hoyle, P.C. Hwang and C. Litch-
    field. 1968. Positional distribution of fatty acids
    in depot fat of aquatic animals. Lipids 3:24-29.

85. Brockerhoff, H. and M. Yurkowski. 1966. Stereospecific
    analysis of several vegetable oils. J. Lipid Res.
    7:62-64.

86. Weber, E.J., I.A. de la Roche and D.E. Alexander. 1971.
    Stereospecific analysis of maize triglycerides.
    Lipids 6:525-530.

87. de la Roche, I.A., E.J. Weber and D.E. Alexander. 1971. Effects of fatty acid concentration and positional specificity on maize triglyceride structure. Lipids 6:531-536.

88. Sahasrabudhe, M.R. and I.G. Farn. 1964. EFfect of heat on triglycerides of corn oil. J. Amer. Oil Chem. Soc. 41:264-267.

89. Raghuveer, K.G. and E.G. Hammond. 1967. The influence of glyceride structure on the rate of autoxidation. J. Amer. Oil Chem. Soc. 44:239-243.

90. Catalano, M., M. de Felice and V. Sciancalepore. 1975. Autoxidation of monounsaturated triglycerides. Influence of the fatty acid position. Ind. Aliment. 14:89-92.

91. Drozdowski, B. 1977. Effect of the unsaturated acyl position in triglycerides on the hydrogenation rate. J. Amer. Oil Chem. Soc. 54:600-603.

92. Raghavan, S.S. and J. Ganguly. 1969. Studies on the positional integrity of glyceride fatty acids during digestion and absorption in rats. Biochem. J. 113:81-87.

93. Simon, E.W. 1974. Phospholipids and plant membrane permeability. New Phytologist 73:377-420.

94. Gubbels, G.H. 1974. Growth of corn seedlings under low temperature as affected by genotype, seed size, total oil and fatty acid content of the seed. Can. J. Plant Sci. 54:425-426.

95. de Silva, N.S., P. Weinberger, M. Kates and I.A. de la Roche. 1975. Comparative changes in hardiness and lipid composition in two near-isogenic lines of wheat (spring and winter) grown at 2°C and 24°C. Can. J. Bot. 53:1899-1905.

96. Thorsteinson, A.J. and J.K. Nayar. 1963. Plant phospholipids as feeding stimulants for grasshoppers. Can. J. Zool. 41:931-935.

97. Wu, G.-S., R.A. Stein and J.F. Mead. 1979. Autoxidation
    of fatty acid monolayers adsorbed on silica gel. IV.
    Effects of antioxidants. Lipids 14:644-650.

98. Coe, E.H., Jr. and M.G. Neuffer. 1977. The genetics of
    corn. In: Corn and Corn Improvement. (G.F. Sprague,
    ed.) Amer. Soc. of Agronomy, Inc. (Madison, WI), pp.
    111-223.

Chapter Six

CHEMISTRY AND BREEDING OF CRUCIFEROUS VEGETABLES

PAUL H. WILLIAMS

Department of Plant Pathology
University of Wisconsin
Madison, WI   53706

INTRODUCTION

The Cruciferae are represented by a number of genera
and species of economic importance as crops, ornamentals
and weeds. Among the crucifers, brassicas, Brassica sp.
and radishes Raphanus sativus are of considerable importance as vegetables, sources of edible and industrial oils,
animal feeds, green manure and condiments.[21] Among the
brassicas are six interrelated species, three of which are
diploid, B. nigra (n = 8), B. oleracea (n = 9), B.
campestris (n = 10) and three of which represent the amphidiploid derivatives of the diploids, B. carinata (n = 17),
B. juncea (n = 18) and B. napus (n = 19) (Yarnell, 1956).
Radish has a haploid genome of 9 chromosomes (n = 9) (Table
1, Figure 1).

140

P. H. WILLIAMS

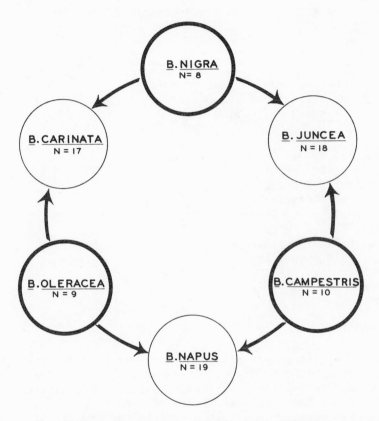

Figure 1.   Interrelationships among three diploid species
of <u>Brassica</u>, <u>B</u>. <u>nigra</u>, <u>B</u>. <u>oleracea</u> and
<u>B</u>. <u>campestris</u>, with three amphidiploid species
<u>B</u>. <u>carinata</u>, <u>B</u>. <u>juncea</u> and <u>B</u>. <u>napus</u>.[14,17]

Table 1. Genome and cytoplasm designation
of Brassica and Raphanus spp

Species	No.	Genome	Cytoplasm
B. nigra	8	bb	B
B. oleracea	9	cc	C
B. campestris	10	aa	A
B. carinata	17	bbcc	BC
B. juncea	18	aabb	AB
B. napus	19	aacc	AC
R. sativus	9	rr	R

Table 2. Common uses of <u>Brassica</u> and <u>Raphanus</u> spp.

Species	Chromo-some No.	Common Name	Use
<u>B. nigra</u>	8	black mustard	mustard, oil, weed
<u>B. oleracea</u>	9	cole crops	vegetable, fodder
<u>B. campestris</u>	10	turnips rapeseed oriental greens	vegetable, oil, fodder
<u>B. carinata</u>	17	Abyssinian mustard	oil vegetable, fodder
<u>B. juncea</u>	18	leaf mustard mustard	oil vegetable, mustard
<u>B. napus</u>	19	rape kale	oil vegetable, fodder
<u>R. sativus</u>	9	radish dikon	vegetable, fodder oil, green manure

Brassicas and radishes display perhaps the greatest morphologic and economic diversity of any group of crop species. Within most species is a range of morphological taxa which reflects a parallelism in selection by man for a wide diversity of utilization (Table 2). Within the cole crops (B. oleracea) can be found the vegetables, common cabbage (B. o. var capitata) with its white, red and savoy types, cauliflower (B. o. var botrytis), broccoli (B. o. var italica), Brussels sprouts (B. o. var gemmifera), kohl rabi (B. o. var gongylodes), Chinese kale (B. o. var alboglabra), collards (B. o. var sabellica), and kales (B. o. var acephala), and the animal fodders, marrow-stem kale (B. o. var medullosa), and thousand-head kale (B. o. var ramosa). An even greater diversity can be found in B. campestris, represented by turnips (B. c. ssp. rapifera), Chinese cabbage (B. c. ssp. pekinenesis), pak choi (B. c. ssp. chinennsis), the oriental greens represented by B. c. ssp. parachinensis, B. c. ssp. nipposinica and B. c. ssp. perviridis and the oil seed turnip rapes (B. c. ssp. oleifera). Leafy and rooted vegetables, fodder and oilseed forms can also be found in B. juncea, B. napus, and R. sativus (Table 2).

INTERSPECIFIC GENETICS

The interrelationships of the six Brassica spp. has permitted the transfer of genes from one to another through interspecific hybridization. Numerous useful traits such as disease resistance and male sterility have been trans- ferred from one species to another via interspecific hybridization. As an example, cytoplasmic male sterility derived from a cross between B. nigra (B1 bb) and B. oleracea (C cc), which initially recreated the amphidiploid B. carinata (B1 bbcc), was transferred to B. oleracea by repeatedly backcrossing B. oleracea to the cytoplasmic male sterile B. carinata. After a number of back crosses Pearson[16] was able to restore high female fertility and recover the complete B. oleracea genome in the male sterile B. nigra cytoplasm (B1 cc) (Figure 2).

There have been a number of reports of the introgression of clubroot disease (Plasmodiophora brassicae) resistance from turnip (Aa'a') to rutabaga (AC aacc) through backcrossing and repeated selfing to stabi- lize the resistance factors,[9] (Figure 3).

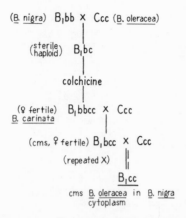

Figure 2.   Substitution of the <u>Brassica oleracea</u> nuclear
genome (cc) into the cytoplasmic male sterile
cytoplasm of <u>B. nigra</u> (B_1) via the synthesis of
a <u>B. carinata</u> (B_1bbcc) bridge.[16]

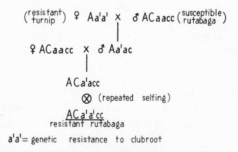

Figure 3.   Interspecific introgression of resistance to
<u>Plasmodiophora brassicae</u> from turnip (Aa'a') to
rutabaga (AC aacc).[9]

INTERGENERIC GENETICS

The classic work of Karpechenko[8] demonstrated the potential of crossing the more widely related radish with brassicas to produce the new genus Raphanobrassica. Recent work by McNaughton at the Scottish Plant Breeding Station[13] has demonstrated the potential of raphanobrassicas as high yielding, multiple disease resistant forages. Bannerot et al.,[1] Heyn[7] and others have illustrated the potential usefulness of Raphanus cytoplasm as a stable source of cytoplasmic male sterility (cms) for various Brassica spp. By crossing cms radish (Rl rr) obtained from Ogura[15] with cabbage (C cc), Bannerot was able to produce a cms Raphanobrassica (Rl rrcc). Repeated backcrosses to cabbage restored female fertility and resulted in the substitution of the cabbage nuclear genome in the cms Rl cytoplasm (Rl cc) (Figure 4).

Crosses of cms B. oleracea (Rl cc) with B. campestris (A aa) permitted the development of cms B. napus (Rl aacc) which then served as a bridge for the production of cms B. campestris (Rl aa) (Figure 5). Similar kinds of crosses have been made to produce cms B. juncea (Rl aabb), B. nigra (Rl bb), and B. carinata (Rl bbcc).

CYTOPLASMIC GENETICS

An important concern when making interspecific or intergeneric crosses that is well illustrated by the work on cytoplasmically inherited male sterility is the need to consider the direction of any particular cross. If cytoplasmic components are likely to play an important role in the desired phenotype, then the choice of the female parent in the cross will be crucial.[23] Figure 3 illustrates this point clearly. In the first cross between turnip and rutabaga the female is turnip bearing the cytoplasm (A) in subsequent crosses, rutabaga with (AC) cytoplasm is selected as the female parent and the normal balance of AC cytoplasm to the aacc nuclear genome is restored. In the case of the Rl cms cytoplasm from radish the integrity of the cytoplasm is maintained by virtue of the fact that no pollen is produced in the Rl substitution lines. Where cms is not involved in the production of intergeneric crosses between Raphanus and various Brassica species, the derived hybrid could be either Raphanobrassica, resulting from a radish

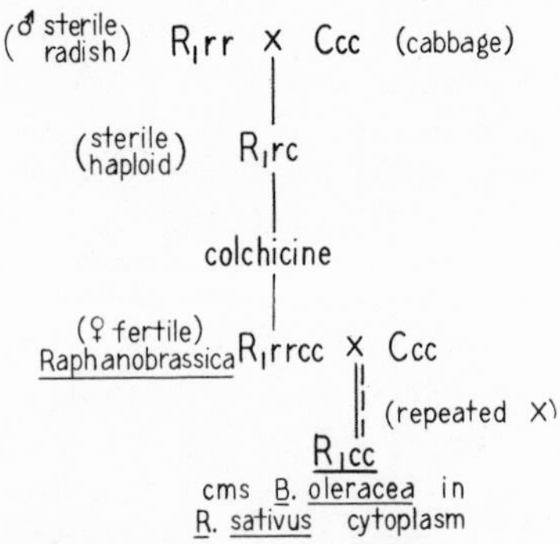

$(\overset{\male \text{ sterile}}{\text{radish}})$ R_1rr X Ccc (cabbage)

$(\overset{\text{sterile}}{\text{haploid}})$ R_1rc

colchicine

$(\overset{\female \text{ fertile}}{\underline{\text{Raphanobrassica}}})$ R_1rrcc X Ccc

(repeated X)

$\underline{R_1cc}$

cms <u>B. oleracea</u> in
<u>R. sativus</u> cytoplasm

Figure 4.   Substitution of the <u>Brassica oleracea</u> nuclear
genome (cc) into the cytoplasmic male sterile
cytoplasm (cms) of <u>Raphanus sativus</u> (R1) via the
synthesis of a <u>Raphanobrassica</u> (R1 rrcc)
bridge.[1]

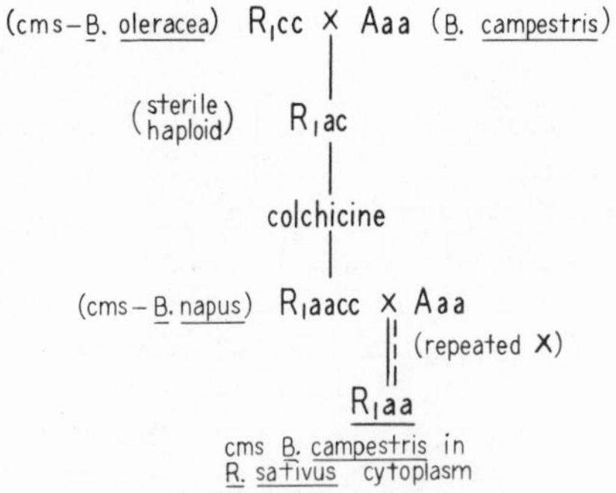

(cms—<u>B</u>. oleracea) R_1cc X Aaa (<u>B</u>. campestris)

$(\overset{\text{sterile}}{\text{haploid}})$ R_1ac

colchicine

(cms—<u>B</u>. napus) R_1aacc X Aaa

(repeated X)

$\underline{R_1aa}$

cms <u>B. campestris</u> in
<u>R. sativus</u> cytoplasm

Figure 5.   Substitution of <u>Brassica campestris</u> nuclear
genome (aa) into the cytoplasmic male sterile
cytoplasm (cms) of <u>Raphanus sativus</u> (R1) via the
synthesis of a <u>B</u>. <u>napus</u> (R1 aacc) bridge.

ovule being fertilized with brassica pollen or Brassico-
raphanus when a brassica ovule is fertilized by radish
pollen. Although the major activity in crucifer vegetable
breeding has occurred within the morphotypes of each
species, there is a growing recognition by breeders of the
potential for exploiting the diversity within the interrelated species via interspecific and intergeneric hybridization.

THE IMPORTANCE OF GLUCOSINOLATES

       An increasingly important aspect of cruciferous
vegetable breeding that has long been recognized by oilseed
and forage crucifer breeders is the need to examine the
chemical constituents of the crucifer cultivars and
breeding lines as an important component of cultivar and
crop phenotype. The crucifers are known to be a rich
source of the glucosinolates, a class of glucose and
sulfur-containing compounds whose enzyme breakdown products
are known for their pungency, their biological activity in
pest attraction and disease resistance, and their role in
certain mammalian metabolic disorders.[18] Glucosinolates
(GS) are characterized by the structure depicted in Figure
6, and are differentiated by the chemical structure of
their aglucone (R) grouping. About 80 different glucosinolates have been isolated from a wide range of species in
the Cruciferae. All crucifers examined have been found to
contain one or more glucosinolates. Upon disruption of
crucifer cells, thioglucosidase glucohydrolase,
(EC 3.2.3.1.), commonly known as thioglucosidase or
myrosinase, acts hydrolytically upon glucosinolates to
release glucose and sulfate ions. The aglucone is usually
converted to isothiocyanate through a Lossen rearrangement,
to nitriles or to thiocyanate (Figure 6).

       Much of the characteristic flavor of cruciferous vege-
tables comes from the isothiocyanates or so-called mustard
oils derived from the glucosinolates. Allylisothiocynate
is the volatile mustard oil responsible for the sharp or
pungent flavor in the condiment mustards, B. nigra and B.
juncea and is also responsible for much of the mustard-like
components of flavor in cabbage, collards, kales and
mustard greens.[18] Studies of the edible parts of commercial cabbages by Van Etten et al.[4,19,20] have identified
and quantified 11 common glucosinolates (GS) of which the

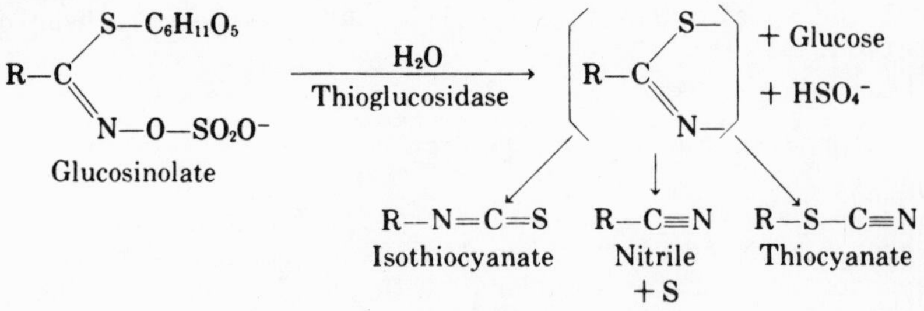

Figure 6. Reaction products of thioglucosidase acting upon a glucosinolate.

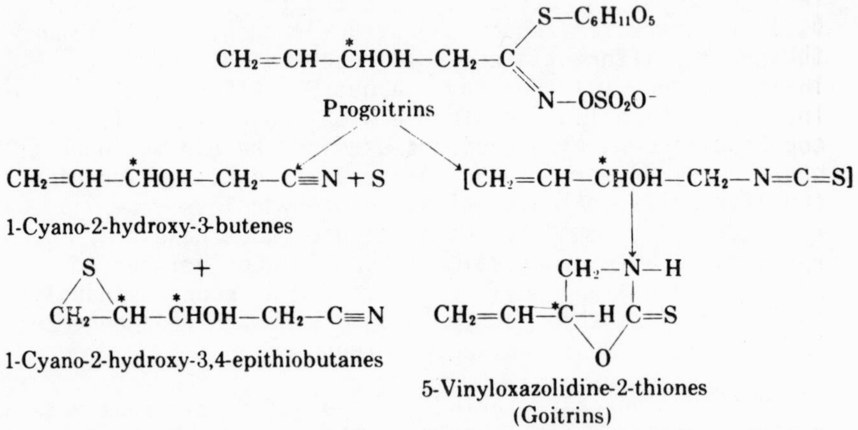

Figure 7. Enxyme hydrolysis products of progoitrin.

major components were 3-methysulfinyl-propyl GS ((glucoiberin), 4-methylsulfinylbutyl GS (glucoraphanin), allyl GS
(sinigrin) and two 3-indolymethyl GS (glucobrassicin and
neoglucobrassicin). An important glucosinolate found in
many brassicas is (r)-2-hydroxy-3-butenyl GS or progoitrin.
The isothiocyanate of progoitrin undergoes cyclization to
form 5-vinyloxazolidine-2-thione or goitrin. Under certain
conditions of enzymatic hydrolysis progoitrin may form an
unsaturated nitrile and two diasteriomeric epithio nitriles
instead of goitrins (Figure 7). Other important glucosinolates are the indolylmethyl GS, glucobrassicin and
neoglucobrassicin which upon hydrolysis yield 3-hydroxy-
methylindoles and thiocyanate ion. The 3-hydroxymethyl-
indoles may react with ascorbic acid to form ascorbigen.
The indolymethylglucosinolates have also been implicated in
the biosynthesis of the plant growth hormone indolylacetic
acid via indolylacetonitrile.[2]

Goitrin together with allylisothiocyanate and thiocyanate ion have been shown to have goitrogenic activity in
test animals. Among the antithiroid agents found in the
Cruciferae, goitrins are the most active.[18] Progoitrin is
considered to be a minor component of the glucosinolates of
most brassicas with the exception of B. napus where it is
a major component of both rutabaga roots and oilseed meal.

GLUCOSINOLATES IN COLE CROPS

Recent surveys by Van Etten and his colleagues at the
U.S.D.A. Northern Regional Research Laboratory, on the
major cultivars of cruciferous vegetables grown in the
United States have indicated a great diversity in the
various glucosinolates within and among vegetable species.
As noted above, the four main glucosinolates found in
cabbage were glucoiberin, glucoraphanin, sinigrin and
glucobrassicin.[20] Correlation analysis of the quantities
of the various glucosinolates from 22 F1 hybrid and open
pollinated cultivars identified four cultivar groupings.
Groups A, B, and C contained more total glucosinolates from
3-carbon (propyl) than from 4-carbon (butyl) aglucones,
whereas in group D, 4-carbon aglucones predominated. Group
C differed in that it was very high in the 3-carbon allyl

glucosinolate but low in the 3- and 4-carbon methyl-
sulfinyl glucosinolates. Groups A and B have similar
patterns but within group A the correlations are slightly
higher than within group B. Inspection of the pedigrees of
the hybrids and the progenitors from which they were
derived indicate a heritable component to their glucosinolate content and points to the need for more critical
studies on the genetics and inheritance of glucosinolates.

Unpublished studies on red, white and savoy cabbages
indicate highly significant differences in the major glucosinolates among the three types of cabbage as well as differences in the distribution in glucosinolates in different
parts of the head. The region of the cambium and cortex
surrounding the pithy core of the cabbage contains about
twice the amount of glucosinolates compared to the pith or
head leaves. Furthermore, the amounts of the various major
components varied significantly between core, cambium and
leaf regions of the head.[19]

Analysis of the various morphological varieties of B.
oleracea indicates distinctly different quantities of
various glucosinolates among broccoli, cauliflower,
Brussels sprouts, collards, kale and kohl rabi. The main
glucosinolate in Brussels sprouts is glucobrassicin
whereas in kohl rabi 4-methylthiobutyl GS (glucoerucin)
predominates. Collards are considerably higher in the
goitrogenic glucosinolates sinigrin and progoitrin than
other cole crops (unpublished data). Glucoraphanin is present as the major component in broccoli whereas it was not
present in the heads of the major cauliflower cultivars
grown in the United States. An interesting aspect illus-
trating the value of knowing the glucosinolate profiles of
various crucifers occurred during our survey of glucosinolates in U.S. vegetable crucifers. Upon analysis, the
cauliflower 'Royal Purple' yielded 95 ppm glucoraphanin and
had a glucosinolate profile much more characteristic of
broccoli than of cauliflower. Upon investigating the origins of Royal Purple and the microscopic floral anatomy, I
was able to determine that it was indeed a form of heading
broccoli which is marketed under the name of cauliflower.
Considerably more variation in the glucosinolate profiles
exist among the kales than in other cole crops. This is
undoubtedly due to the rather diverse origins of a number
of cultivars that are classified as kale.

GLUCOSINOLATES IN CHINESE CABBAGE

Analysis of the glucosinolates from 14 hybrid and open pollinated cultivars of Chinese cabbage, (B. campestris ssp. pekinensis) indicate that glucosinolates yielding 5-carbon aglucons predominate whereas 3- and 4-carbon aglucons predominate in common cabbage.[5] As was found for cabbage, the quantities of individual glucosinolates in Chinese cabbage varied considerably among cultivars. Also, as has been generally observed for crucifers, both the seeds and younger tissues of Chinese cabbage contained greater amounts of glucosinolates than mature leaves.

GLUCOSINOLATES AND HEALTH

The publication in the Federal Register on June 25, 1971 p. 12094 of "Eligibility for classification as generally recognized as safe (GRAS)" of the Food and Drug Administration (FDA) regulation has prompted renewed interest in seeking more information on the level of toxic constituents in Brassica vegetables.[6] Though at this time only a beginning has been made toward investigating the glucosinolates in vegetables, the improved quantitative and qualitative methodologies developed by Daxenbichler and his colleagues[3,4] for glucosinolate analysis of plant tissues will be of considerable value in providing important "base-line" information on the levels and variation of potentially useful and hazardous glucosinolates. Preparative techniques for glucosinolates will also permit biological and toxicological evaluation of specific compounds. Despite the fact that much of the concern over glucosinolates has been generated on the basis of largely unsubstantiated evidence that the consumption of excessive amounts of brassica crops has resulted in benign goiter,[6] there is good reason to believe that by learning more about the heritability and genetics of glucosinolates in crucifers such information may be of considerable use to plant breeders in the future. By understanding the relationship of flavor to the composition and content of glucosinolates, breeders, with appropriate genetic knowledge of glucosinolates may be able to develop cultivars of enhanced quality.[12] Reports on the role of benzylisothiocyanate and phenylethyl isothiocyanate in inhibiting carcinogen-induced neoplasia in rats and mice through the stimulation of mixed function oxidases[22] point to the possibility of introducing into our diets, levels of potentially therapeutic chemical constituents. The beneficial effects of crucifer residues incorporated in soils in reducing the severity of pea root rot (Aphanomyces euteiches) has been attributed to the fungitoxic activity of allyisothiocyanate and other sulfur-containing breakdown products.[10] The insecticidal activity of 2-phenylethyl isothiocyanate from B. oleracea[11] might be more effectively exploited if cultivars containing high levels of the chemical were developed.

As crucifer breeders begin to exploit the diversity within Brassica and Raphanus species through interspecific and intergeneric hybridization, there will be an increasing need to monitor the chemical phenotypes of their resulting crosses. Until now the chemical constituents of most of the cultivars within each vegetable species fall within a spectrum that is representative for that species. As wider crosses are made, the resultant progenies are likely to carry vastly different arrays of glucosinolates than their progenitors. Precisely how these new patterns will be expressed in terms of flavor, palatability, toxicity and biological activity remains to be seen.

SUMMARY

Among the crucifers six interrelated species of brassica and radish constitute important crops in world agriculture. Within each species is a wide range of morphological diversity that satisfies various forms of utilization from vegetables, fodder, condiments to edible and industrial oils. B. juncea, B. carinata and B. napus are amphidiploid progenitors of the diploid B. nigra, B. oleracea and B. campestris. Hybridization among the six interrelated brassicas and radish has permitted the transfer of many important traits such as disease and pest resistance, male sterility and improved yield and quality. An important class of chemicals found in crucifers are the glucosinolates some of which are known for their pungency, for their insect attractiveness and implicated in possible metabolic disorders in animals and man. An important aspect of crucifer breeding programs therefore is the acquisition of qualitative and quantitative information on glucosinolates and on their genetics and heritability. Recent analyses of various crucifer vegetables have indicated the distinctiveness of numerous species-related glucosinolates. Studies of hybrids and their parents indicate the heritability of glucosinolates and points to the potential for transfer of specific glucosinolate biosynthesis from one species to another.

ACKNOWLEDGEMENTS

Research supported by the College of Agricultural and Life Sciences, University of Wisconsin, Madison, by gifts from the National Kraut Packers Association, and by USDA-SEA Hatch formula funds Project No. 2378. The author is indebted to C. H. Van Etten and M. E. Daxenbichler, USDA-SEA, NRRL, Peoria, Ill., for their collaborations and to S. A. Vicen for assistance in the preparation of the figures.

REFERENCES

1. Bannerot, H., L. Loulidard, Y. Cauderon and J. Tempe. 1974. Transfer of cytoplasmic male sterility from Raphanus sativus to Brassica oleracea, pp. 52-54. In: Cruciferae, 1974. Proc. Meeting Veg. Crops. Sec., Eucarpia. Sept. 1974.

2. Butcher, D.N., S. El-Tigani and D.S. Ingram. 1974. The role of indole glucosinolates in the club root disease of the Cruciferae. Physiol. Plant Path. 4:127-141.

3. Daxenbichler, M.E. and C.H. Van Etten. 1977. Glucosinolates and derived products in cruciferous vegetables. Gas-liquid chromatographic determination of the aglucone derivatives from cabbage. J. Assn. Off. Anal. Chem. 60:950-953.

4. Daxenbichler, M.E., C.H. Van Etten and G.F. Spencer. 1977. Glucosinolates and derived products in cruciferous vegetables. Identification of organic nitriles from cabbage. J. Agr. Food Chem. 25:121-124.

5. Daxenbichler, M.E., C.H. Van Etten and P.H. Williams. 1979. Glucosinolates and derived products from cruciferous vegetables. Analysis of 14 varieties of Chinese cabbage. J. Agr. Food Chem. 27:34-37.

6. Hanson, C.H. (ed.) 1974. The effect of FDA Regulations (GRAS) on plant breeding and processing. Spec. Pub. No. 5. Crop Sci. Soc. of America, Madison, WI, 63 pp.

7. Heyn, F.W. 1976. Transfer of restorer genes from
   Raphanus to cytoplasmic male sterile Brassica napus.
   Eucarpia, Cruciferae Newsletter 1:15-16.

8. Karpechenko, G.D. 1924. Hybrids of Raphanus sativus L.
   X Brassica oleracea L. J. of Genetics 14:375-396.

9. Lammerink, J. 1970. Interspecific transfer of clubroot
   resistance from Brassica campestris L. to B. napus L.
   New Zealand J. Agr. Res. 13:105-110.

10. Lewis, J.A. and G.C. Papavizas. 1971. Effect of sulfur-
    containing volatile compounds and vapors from cabbage
    decomposition on Aphanomyces euteiches of peas.
    Phytopathology 61:208-214.

11. Lichtenstein, E.P., D.G. Morgan and C.H. Mueller. 1964.
    Naturally occurring insecticides in cruciferous
    crops. J. Agr. Food Chem. 12:158-161.

12. MacLean, A.J. 1976. Volatile flavour compounds of the
    cruciferae, pp. 307-330. In: (J.G. Vaughan et al.,
    eds.) The Biology and Chemistry of the Cruciferae.
    Academic Press (London), 355 pp.

13. McNaughton, I.H. and C.L. Ross. 1978. Inter-specific
    and inter-generic hybridization in the Brassicae with
    special emphasis on the improvement of forage crops.
    Scottish Plant Breeding Stat. Ann. Rept. 57:75-110.

14. Morinaga, T. 1934. Interspecific hybridization in
    Brassica VI. The cytology of F_1 hybrids of B. juncea
    and B. nigra. Cytologia 6:62-67.

15. Ogura, H. 1968. Studies on the new male sterility in
    Japanese radish with special reference to utilization
    of this sterility toward the practical raising of
    hybrid seeds. Mem. Fac. Agr. Kagoshima Univ. 6:39-78.

16. Pearson, O.H. 1972. Cytoplasmically inherited male
    sterility characters and flavor components from the
    species cross Brassica nigra (L.) Koch X B. oleracea
    L. J. Amer. Soc. Hort. Sci. 97:397-402.

17. U. N. 1935. Genome-analysis in Brassica with special reference to the experimental formation of B. napus and peculiar mode of fertilization. Japan J. Bot. 7:389-452.

18. Van Etten, C.H. and I.A. Wolff. 1973. Natural sulfur compounds, pp. 210-234. In: Toxicants Occurring Naturally in Foods (2nd ed.) Committee on Food Protection, Nat'l. Acad. Sci., Washington, D.C.

19. Van Etten, C.H., M.E. Daxenbichler, W.F. Kwolek and P.H. Williams. 1979. Glucosinolates and derived products in crruciferous vegetables: Distribution of glucosinolates in the pith, cambial - cortex, and leaves of the head in cabbage, Brassica oleracea L. J. Agr. Food Chem. 27 (in press).

20. Van Etten, C.H., M.E. Daxenbichler, P.H. Williams and W.F. Kwolek. 1976. Glucosinolates and derived products in cruciferous vegetables. Analysis of the edible part in twenty-two varieties of cabbage. J. Agr. Food Chem. 24:452-455.

21. Vaughan, J.G., A.J. MacLeod and B.M.G. Jones (eds.) 1976. The Biology and Chemistry of the Cruciferae. Academic Press (London), 355 pp.

22. Wattenberg, L.W. 1978. Inhibition of chemical carcinogenesis. J. Nat'l. Cancer Inst. 60:11-18.

23. Yarnell, S.H. 1956. Cytogenetics of vegetable crops. II Crucifers. Bot. Rev. 22:81-166.

Chapter Seven

CHEMICAL GERMPLASM INVESTIGATIONS IN SOYBEANS:
THE FLOTSAM HYPOTHESIS

THEODORE HYMOWITZ

Department of Agronomy
University of Illinois
Urbana, IL   61801   USA

Introduction
    Origin and domestication of the soybean
    Chemical compositional changes in seed
        as a consequence of domestication
    Flotsam hypothesis
Kunitz trypsin inhibitor
Seed lectins
Soybean Amylases
Summary
References

INTRODUCTION

Origin and domestication of the soybean

    The genus Glycine Willd. is currently divided into two
subgenera Glycine and Soja (Moench) F. J. Herm.[1] The subgenus Soja includes the soybean, G. max (L.) Merr. and its
closest relative, the wild soybean, G. soja Sieb. and Zucc.
(Table 1.). The wild soybean is found in open fields,
hedgerows, along roadsides and riverbanks in the Republic
of China, adjacent areas of the USSR (Primorskiy and
Khaborovsk Kray), Korea, Japan and Taiwan.[2] Both G. max
and G. soja are diploids (2n=40).[3-10] Evidence gathered
from cytogenetic, morphological and seed protein studies
suggest that the two species are conspecific[11-16] and
supports the hypothesis that G. soja is the wild ancestor
of the cultivated soybean.[3,4] There are few, if any,
cytogenetic barriers to hybridization between the two
species. The major difference between G. max and G. soja

157

Table 1. Chromosome Number and Geographic Distribution
of Species in the Genus Glycine

	Species	2n	Distribution
	Subgenus GLYCINE		
1.	G. clandestina Wendl.	40	Australia; South Pacific Islands
1.a	var sericea Benth.	--	Australia
2.	G. falcata Benth.	40	Australia
3.	G. latrobeana (Meissn.) Benth.	40	Australia
4.	G. canescens F.J. Herm	40	Australia
5.	G. tabacina (Labill.) Benth.	40,80	Australia; South China; Taiwan; South Pacific Islands
6.	G. tomentella Hayata	38,40,78,80	Australia, South China; Taiwan; Philippines; Papua New Guinea
	Subgenus SOJA (Moench) F.J. Herm		
7.	G. soja Sieb & Zucc.	40	China; Taiwan; Japan; Korea; U.S.S.R.
8.	G. max (L.) Merr.	40	Cultigen

are properties expected between a domesticate and its wild
ancestor. The wild soybean is a sprawling or climbing vine
with slender, many branched stems and small oval to oblong
black seeds weighing from about 10 to 30 mg each.[2] On the
other hand, the cultivated soybean is an erect, sparsely
branched herb with large ovoid to sub-spherical yellow,
green, black, brown or reddish brown seeds weighing from
100 to 400 mgm each.[2] Historical, linguistic and geographical evidence point to the eastern half of North China as
the region where the soybean first emerged as a domesticate
around the 11th century B.C.[17] Today the soybean is widely
planted in both the Old and New Worlds.

Chemical compositional changes in soybean seeds
      as a consequence of domestication

    In addition to the above mentioned agronomic changes
that took place on domestication of the soybean, there were
accompanying chemical compositional changes in the seed.
Two of the most significant changes were (a) a major
increase in the total oil content and (b) a slight decrease
in the total protein content in seed. The total oil
content in wild soybean seed is about 10% whereas in the
cultivated soybean seed the total oil content is around
21%.[18] The doubling of the oil content in the seed did
not affect equally all of the major fatty acids, that is,
palmitic, stearic, oleic, linoleic and linolenic. Only
oleic and linoleic acids substantially increased in
concentration in cultivated soybean seeds, while the other
fatty acids are found at concentrations similar to those in
wild soybean seed[1,18] (Table 2). The slight decrease in
the total protein content in seed of cultivated soybeans
(about 41%) compared to wild soybeans (about 45%) is not
unique. A similar phenomenon has been reported in many
other wild species--cultigen complexes.[19,20] The quantitative shifts in the oil and protein content in the seed
are a consequence of selection by the farmers of China for
soybean plants with higher yields. Modern soybean breeders
have determined that there is a positive association between oil content in the seed and yield, whereas there is a
negative association between protein content and yield.[21]

Flotsam hypothesis

    Another major consequence of the soybean domestication
process was the transmission from G. soja to G. max of

Table 2. Oil Content and Fatty Acid Distribution
in the Genus Glycine

	Oil (%)	Weight of Fatty Acid in 100 g Seed*				
		P	S	O	Li	Le
G. clandestina	13.9	2.7	0.9	3.1	5.2	2.0
G. falcata	13.7	2.7	1.7	2.3	5.0	2.0
G. canescens	12.7	2.4	0.6	1.3	5.4	3.0
G. tabacina	11.2	2.5	0.5	1.8	4.6	1.7
G. tomentella	12.8	2.7	0.8	2.2	4.9	2.1
G. soja	9.8	1.4	0.3	1.3	5.2	1.5
G. max	21.0	3.2	0.8	5.0	10.3	1.7

* The product of the mean percentage of each fatty acid and
the mean percentage oil content of each species. P =
palmitic, S = steric, O = oleic, Li = linoleic and Le =
linolenic.

certain chemical components of seed. These chemical
components apparently play no significant role in the
structure or metabolism of the wild or cultivated soybean.
Perhaps these chemical components of seed originally
evolved in G. soja as defensive mechanisms against
pathogens, insects and vertebrates, as protective agents
against fluctuating climate conditions, or as a general
strategy to compete against other plants. The wild soybean
is an annual that produces seed each fall which must
survive a harsh winter and then germinate the following
spring. On the other hand, G. max is a species nurtured by
man. Each fall, the farmer harvests his soybean crop and
stores the future planting seed under the most favorable
conditions. In the spring, the farmer prepares and amends
the land (plows, puts on fertilizers, herbicides, irri-
gates, etc.) and then plants the seed at the proper time to
ensure good growth and a successful yield. The natural
selective forces acting in the wild are buffered by man
from acting upon the cultivated soybean. The transmission
of these defensive or protective chemical agents from the
wild soybean where they ensure the survival of the species
to the cultivate soybean can be explained by the "Flotsam
Hypothesis." This attempts to explain the presence of
certain chemical components in cultivated soybean seed that
apparently have little or no major role in the survival of
the species. In addition, the hypothesis suggests new
approaches to breeding soybeans for protein with altered
functional properties, for improving the nutritional
quality of the soybean, and perhaps for increasing yields.
Possible examples of flotsam in soybean seed are the Kunitz
trypsin inhibitor, seed lectins, and β-amylase.

KUNITZ TRYPSIN INHIBITOR

More than 60 years ago, Osborne and Mendel[22] demonstrated that unheated soybean meal is inferior in nutritional quality to properly heated soybean meal. The
trypsin inhibitor proteins in raw mature soybean seed have
been proposed to be one of the major factors responsible
for the poor nutritional value.[23,24] Ingestion of raw
soybeans causes pancreatic hypertrophy[29-32] and is believed
to inhibit growth by upsetting the balance between methionine and cystine in the pancreas.[31,33] Several different
trypsin inhibitors have been reported to be present in
soybeans.[25,34,-39] However, much of the soybean trypsin

inhibitor A_2 (SBTI-A_2)[26] which was first crystallized by Kunitz[37] and commonly is known as the Kunitz trypsin inhibitor. The protein contains 181 amino acids and has a molecular weight of 21,384.[40]

Seed from the U.S. Department of Agriculture Glycine germplasm collection (2944 accessions of G. max and 361 accessions of G. soja) were screened using polyacrylamide gel electrophoresis for the presence or absence of electrophoretic forms of SBTI-A_2. Thus far, four electrophoretic forms of SBTI-A_2 have been discovered in the G. max germplasm (Figure 1). Three of the forms designated Ti[a], Ti[b] and Ti[c] are electrophoretically distinguishable from one another by their different R_p values of 0.79, 0.75 and 0.83, respectively[41,47,48] (R_p is the mobility relative to a bromophenol blue dye front in a 10% polyacrylamide gel anodic system in which a Tris-glycine buffer, pH 8.3, is used). The three forms are controlled by a codominant multiple allelic system at a single locus.[49,50] In the G. soja germplasm only Ti[a] and Ti[b] were found.[42] The fourth form of SBTI-A_2 found in G. max germplasm (Plant Introductions 157440 and 196168) does not exhibit a protein band in the gels.[42] The lack of a protein band is inherited as a recessive allele and has been designated ti.[48] In the homozygous recessive state (ti ti) no SBTI-A_2 band is produced in the seed. A standard trypsin assay of crude protein extracts from seeds which lacked the SBTI-A_2 revealed that the extracts had 30 to 50 percent less trypsin inhibitor activity per gram protein than crude protein extracts from seeds of the cultivar Amsoy 71 which contains SBTI-A_2 (Ti[a]). Although other trypsin inhibitors may be present in seed of Plant Introductions 157440 and 196168, all experimental data indicate that the Kunitz trypsin inhibitor is lacking in those seeds which do not have the SBTI-A_2 protein.

The significance of the Kunitz trypsin inhibitor in cultivated soybean seed has not been established. It would appear that the protein does not function as a regulator of metabolism as the seed of the two plant introductions without the Kunitz trypsin inhibitor germinate normally and the subsequent plants grow, flower and set seed just like typical soybean seed that have the inhibitor. Most probably the Kunitz trypsin inhibitor evolved in the genus Glycine as a defensive mechanism or as a protective agent against soil pathogens or herbivores.[51] An interesting

Figure 1. Polyacrylamide gels of protein extracts from seeds showing SBTI-A$_2$ bands. From left to right:

1) no SBTI-A$_2$ band (<u>ti</u>);

2) R$_p$ 0.75 (<u>Ti</u>^{b});

3) R$_p$ 0.79 (<u>Ti</u>^{a});

4) R$_p$ 0.83 (<u>Ti</u>^{c}).

possibility as to why the Kunitz trypsin inhibitor has been
maintained in soybean seed after several millenia of culti-
vation may be due to its linkage with one of the acid
phosphatases, enzymes that have a significant role in plant
metabolism.[52]

SEED LECTINS

Among the biologically active components of cultivated
soybean seeds are a group of glycoproteins that cause the
agglutination of certain red blood cells. These glycoproteins are called lectins or phytohemagglutinins.[32,53]
The presence of lectins in soybean seed was first reported
in 1909.[54] However, it was not until 1952 that a lectin
was isolated and purified.[55] At least four different
lectin forms have been reported in soybean seed.[55-60]
Defatted soybean meal contains about 3% lectins.[61] The
major lectin present in the seed of most soybeans has a
molecular weight of 120,000 daltons and is composed of four
subunits.[56,62,63] The lectin has specificity for N-acetyl-
D-glactosamine and to a lesser extent for D-galactose.[64]

Seed from the U.S. Department of Agriculture Glycine
germplasm collection were screened using polyacrylamide gel
electrophoresis for the presence or absence of electrophoretic forms of soybean seed lectin (SBL).[65-67] Five
soybean lines (Columbia, Norredo, Sooty, T-102, and
Wilson-5) were found that lacked the seed lectin.[68] In
addition to polyacrylamide gel electrophoresis, the absence
of lectin was confirmed by affinity chromatography, hemag-
glutination, and binding of cells by certain rhizobial
strains. Of the 97 soybean lines tested that had seed
lectin, the lectin content varied dramatically. The values
ranged from 2.5 to 12.2 mg SBL per gram of defatted meal.[67]
Thus far, two electrophoretic forms of SBL have been discovered in the G. max germplasm. One form designated Le has
a SBL band at R_p 0.48 (R_p is the mobility relative to the
lysozyme protein band in 10% polyacrylamide gel cathodic
system in which a glycine-citric acid buffer, pH 3.7, is
used)[69] (Figure 2). The second form does not exhibit a
SBL band in the gels. The lack of a SBL band is inherited
as a recessive allele (le). In the homozygous recessive
state (le le) no lectin is produced in the seed.

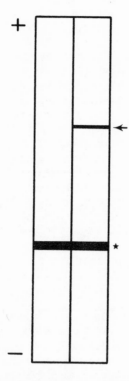

Figure 2. Polyacrylamide gels of protein extracts from seeds
showing the presence or absence of the SBL band.
From left to right: no SBL band (1e) and SBL band
at R_p 0.48 (Le). Arrow points to the SBL band.
Asterisk indicates the lysozyme reference band.

Although the significance of seed lectins has not been established, several suggestions have been made as to their possible function. Included in these suggestions are: (a) lectins may be one reason for the poor nutritional value of raw soybean meal;[70] (b) lectins may play a role in the recognition and binding of rhizobia to the surface of the plant root;[71-73] (c) lectins may provide membrane recognition sites that are involved in the regulation of all division, cell growth or cell differentiation;[74-75] and (d) lectins may function as protective inhibitors of potential pathogens or pests.[51,76-78] It appears that the glycoprotein does not function as a regulator of cell division, growth or differentiation as the soybean lines without SBL germinate normally and the subsequent plants grow, flower and set seed just like typical soybean seed that have SBL. Recent studies have shown that SBL does not appreciably contribute to the poor nutritional value of raw soybean meal,[79,80] that SBL is not linked to the Kunitz trypsin inhibitor,[81] and that soybeans without SBL nodulate as well as soybeans containing seed lectin.[68] Most probably SBL evolved in the genus Glycine as a defensive mechanism or as a protective agent against potential pathogens or pests.

SOYBEAN AMYLASES

β-Amylase is a biologically active component of the 7S protein fraction of cultivated soybean seed.[82] Mature soybean seeds contain both α and β-amylase.[83-85] The β-amylase activity in soybean seeds is much higher than α-form.[83,84,46] Soybean β-amylase was first crystallized by Fukumoto and Tsujisaka;[84] it is a simple protein having a molecular weight of 57,000 daltons and consists of 494 amino acid residues.[88]

Gorman and Kiang,[89,90] utilizing polyacrylamide slab gel electrophoresis, observed cultivar-specific electrophoretic zymograms for amylase (AM) activity. One of the migrating bands (designated AM-3) had two variants, slow and fast AM. In addition, they found two cultivars without AM activity. The first designated AM null-1 was found in some seeds of the cultivar Altona and the second designated AM null-2 was found in all seeds tested of the cultivar Chestnut (Figure 3).

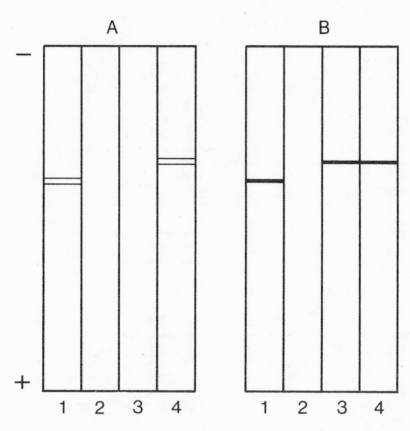

Fig. 3. Polyacrylamide gels of protein extracts from seeds showing (A) the presence or absence of amylase activity and (B) the $\underline{Sp_1}$ bands. (A) Gels stained for amylase after electrophoresing for 45 min. Left to right: 1) Altona, fast AM band; 2) Altona, no AM band; 3) Chestnut, no AM band; and 4) Amsoy 71, slow AM band. (B) The same gels after staining for protein. Left to right: 1) Altona, $\underline{Sp_1}^b$; 2) Altona, no $\underline{Sp_1}$ band; 3) Chestnut, $\underline{Sp_1}^a$; and 4) Amsoy 71, $\underline{Sp_1}^a$.

Hildebrand and Hymowitz[91] demonstrated that the variant
AM band of Gorman and Kiang is β-amylase. In addition,
they reported that the slow AM and fast AM forms of the
variant AM-3 band is the same protein as the electrophoretic variant seed protein designated 'A' and 'B'
respectively by Larsen.[92] Larsen and Caldwell reported
that the 'A' and 'B' seed protein bands are controlled by
codominant alleles at a single locus.[93] Orf and
Hymowitz[94] using polyacrylamide disc electrophoresis
established that the soybean seed band called 'A' by Larsen
occurs at R_p 0.36 and the seed protein band called 'B'
occurs at R_p 0.42 (R_p here is the mobility relative to a
bromophenol blue dye front in a 10% polyacrylamide gel
anodic system in which a Tris-glycine buffer, pH 8.3, is
used). They proposed the gene symbols $\underline{Sp_1}^a$ and $\underline{Sp_1}^b$ for
the electrophoretic forms that occur at R_p 0.36 and 0.42
respectively. Hildebrand and Hymowitz[91] also reported that
the Altona genotype that lacked the AM-3 band (AM null-1)
also lacked ther $\underline{Sp_1}$ seed protein band, the Altona genotype
with normal AM activity produced the $\underline{Sp_1}^b$ seed protein
band, and Chestnut, although it had no detectable AM-3 band
(AM null-2) produced the $\underline{Sp_1}^a$ seed protein band (Figure 3).

Inheritance studies by Hildebrand and Hymowitz[95]
revealed that $\underline{Sp_1}^a$, $\underline{Sp_1}^b$, AM null-1, and AM null-2 form a
multiple allelic series at a single locus. $\underline{Sp_1}^a$ and $\underline{Sp_1}^b$
are codominant alleles for seed protein bands and AM
activity. AM null-1 is recessive to $\underline{Sp_1}^a$ and $\underline{Sp_1}^b$ for seed
protein bands and AM activity. AM null-2 is codominant to
$\underline{Sp_1}^b$ for seed protein band and recessive to $\underline{Sp_1}^a$ and $\underline{Sp_1}^b$
for AM activity. The investigators proposed the gene
symbol $\underline{Sp_1}$ for the AM null-1 allele and $\underline{Sp_1}^{an}$ for the AM
null-2 allele.

Seed from the U.S. Department of Agriculture $\underline{Glycine}$
germplasm collection were screened for variants of the
Sp_1 protein.[95] Of the 2979 accessions of $\underline{G.\ max}$ tested,
89% had the $\underline{Sp_1}^b$ allele. Of the 359 accessions of $\underline{G.\ soja}$
tested 64% had the $\underline{Sp_1}^b$ allele.

Mature soybean seeds contain substantial β-amylase
activity[86] yet they contain only about 1% starch.[97,98]
According to Yazdi-Samadi $\underline{et\ al}$.[98] starch reached a maximum
value of 14.6 mg per seed in the cultivar Harosoy 63 and
7.9 mg per seed in the cultivar Steele at 40 and 30 days
after flowering, respectively. Starch then declined

sharply in both cultivars. On the other hand, Birk and Waldman[86] found that amylase activity appears to increase in soybean seed until maturity. According to Dunn,[99] "α-amylase is the only degradative enzyme which has any action on the starch granule _in vivo_" and the function of the other starch degrading enzymes such as β-amylase is to degrade dextrins released into solution after the α-amylolytic attack on the granule itself. Soybean seed contains alternative enzymes such as starch phosphorylase that perhaps can replace β-amylase. Whether Altona genotype (AM null-1) and Chestnut (AM null-2) are lacking or have extremely low β-amylase activity has not yet been fully established. Nevertheless the soybean lines lacking β-amylase activity germinate normally and the subsequent plants grow, flower and set seed just like typical soybean seed that have normal β-amylase activity.

SUMMARY

       Proposed roles for secondary compounds in seed range from vital to no apparent function. Ecologists[100-102] generally champion the view that the secondary compounds are defensive or protective agents against pests, pathogens or plant competitors. On the other hand, phytochemists such as Seigler[103] maintain that "secondary compounds in plants exist in a state of dynamic equilibrium and are not static end products." Perhaps they have a primary role as sources of nitrogen, sulfur, phosphorus, carbon, etc. In this paper the genetic approach for studying the possible role of certain compounds in soybean seed was reviewed.

       Several thousand lines of soybeans from the U.S. Department of Agriculture _Glycine_ germplasm collection were screened, primarily by polyacrylamide gel electrophoresis for the presence or absence of electrophoretic forms of the Kunitz trypsin inhibitor, seed lectin and β-amylase. Nulls for all three components of seed were located in various _G. max_ accessions and the modes of inheritance were elucidated. None of the three components of seed appear to be vital to the species since the null seed for the Kunitz trypsin inhibitor, seed lectin or β-amylase germinate normally and the subsequent plants grow, flower and set seed just like the typical soybean seed.

The "Flotsam hypothesis" has been proposed to account for the presence of chemical components of soybean seed that apparently are not vital for the survival of the cultigen. According to the hypothesis, certain chemical compounds of soybean seed evolved in G. soja, the wild ancestor of the soybean, as defensive or protective agents. These chemicals were transmitted to the soybean as a consequence of the domestication process. Since the soybean plant is nurtured by man, these compounds are no longer needed as defensive or protective agents, and hence are essentially flotsam in soybean seed.

The availability of soybean lines lacking the Kunitz trypsin inhibitor, seed lectin and β-amylase should be of particular interest to the scientific community for developing soybean cultivars with improved nutritional quality, with protein of altered functional characteristics and perhaps for increasing yields. Lastly, the genetic approach used to determine whether certain chemical compounds in soybean seed are vital to the cultigen can be used as a model for establishing the function of chemical compounds in other plant species.

REFERENCES

1. Hymowitz, T., C.A. Newell. (n.d.) In: (R.J. Summerfield, ed.) Advances in Legume Science. Royal Botanic Garden, Kew.

2. Herman, F.J. 1962. A revision of the genus Glycine and its immediate allies. USDA Techn. Bull. 1268:1-79.

3. Hadley, H.H., T. Hymowitz. 1973. In (B.E. Caldwell, ed.) Soybeans: Improvement, Production and Uses. Amer. Soc. of Agron. (Madison, WI) Chapter 3.

4. Fukada, Y. 1933. Cyto-genetical studies on the wild and cultivated Manchurian soybeans (Glycine L.) Jap. J. Bot. 6:489-506.

5. Karasawa, K. 1936. Crossing experiments with Glycine soja and G. ussuriensis. Jap. J. Bot. 8:113-118.

6. Karpenchenko, G.D. 1925. On the chromosomes of the Phaseolinae. Bull. Appl. Bot. and Plant Breeding (Leningrad) 14:143-148.

7. Kawakami, J. 1930. Chromosome numbers in Leguminosae. Bot. Mag., Tokyo, 44:319.

8. Veatch, C. 1934. Chromosomes of the soybean. Bot. Gaz. 96:189.

9. Tschechow, W., N. Kartaschowa. 1932. Karyologisch-systematische untersuchung der tribus Loteae and Phaseoleae. Unterfam. Papilionatae. Cytologia 3:221-249.

10. Newell, C.A., T. Hymowitz. 1978. Seed coat variation in Glycine Willd. subgenus Glycine (Leguminosae) by SEM. Brittonia 30:76-88.

11. Lu, Y.C. 1966. Studies on the morphology, physiology and cytogenetics of cultivated, semi-cultivated and wild soybeans. J. Agr. Forest 15: 1-23.

12. Mies, D.W., T. Hymowitz. 1973. Comparative electro-phoretic studies of trypsin inhibitors in seed of the genus Glycine. Bot. Gaz. 134:121-125.

13. Tang, W.T., C.H. Chen. 1959. Preliminary studies on the hybridization of the cultivated and wild bean [Glycine max Merrill and G. formosana (Hosokawa)]. J. Agr. Ass. China N.S. 28:16-23.

14. Tang, W.T., G. Tai. 1962. Studies on the qualitative and quantitative inheritance of an interspecific cross of soybean, Glycine max x G. formosana. Bot. Bull. Acad. Sinica 3:39-54.

15. Ting, C.L. 1946. Genetic studies on the wild and cultivated soybeans. J. Amer. Soc. Agron. 38:381-393.

16. Weber, C.R. 1950. Inheritance and interrelation of some agronomic and chemical characters in an interspecific cross in soybeans, Glycine max x G. ussuriensis. Iowa Agr. Exp. Sta. Res. Bull. 374:767-816.

17. Hymowitz, T. 1970. On the domestication of the soybean.
    Econ. Bot. 24:408-421.

18. Hymowitz, T., R.G. Palmer, H.H. Hadley. 1972. Seed
    weight, protein, oil and fatty acid relationships
    within the genus Glycine. Trop. Agric. (Trinidad)
    49:245-250.

19. Harlan, J.R. 1967. A wild wheat harvest in Turkey.
    Archaeology 20:197-201.

20. Harlan, J.R. 1976. Genetic resources in wild relatives
    of crops. Crop Sci. 16:329-333.

21. Brim, C.A. 1973. In: (B.E. Caldwell, ed.) Soybeans:
    Improvement, Production and Uses. Amer. Soc. of
    Agron. (Madison, WI), Chapter 5.

22. Osborne, T.B., L.B. Mendel. 1917. The use of soybean as
    feed. J. Biol. Chem. 32:369-377.

23. Borchers, R., C.W. Anderson, F.E. Mussehl, A. Moehl.
    1948. Trypsin inhibitors. VIII. Growth inhibiting
    properties of a soybean trypsin inhibitor. Arch.
    Biochem. 19:317-322.

24. Westfall, R.J., S.M. Hauge. 1948. The nutritive quality
    and the trypsin inhibitor content of soybean flour
    heated at various temperatures. J. Nutr. 35:379-389.

25. Rackis, J.J., R.L. Anderson. 1964. Isolation of four
    soybean trypsin inhibitors by DEAE-cellulose
    chromatography. Biochem. Biophys. Res. Commun.
    15:230-235.

26. Rackis, J.J., H.A. Sasame, R.K. Mann, R.L. Anderson,
    A.K. Smith. 1962. Soybean trypsin inhibitors:
    Isolation, purification and physical properties.
    Arch. Biochem. Biophys. 98:471-478.

27. Rackis, J.J. 1965. Physiological properties of soybean
    trypsin inhibitors and their relationships to
    pancreatic hypertrophy and growth inhibition of rats.
    Feder. Proc. 24:1488-1493.

28. Kakade, M.L., D.E. Hoffa, I.E. Liener. 1973. Contribution of trypsin inhibitors to the deleterious effects of unheated soybeans fed to rats. J. Nutr. 103: 1772-1778.

29. Bray, D.J. 1964. Pancreatic hypertrophy in layering pellets induced by unheated soybean meal. Poultry Sci. 43:382-384.

30. Chernick, S.S., S. Lepkovsky, I.L. Chaikoff. 1948. A dietary factor regulating the enzyme content of the pancreas. Changes induced in size and proteolytic activity of the chick pancreas by the ingestion of raw soybean meal. Am. J. Physiol. 155:33-41.

31. Liener, I.E., M.L. Kakade. 1969. In: (I.E. Liener, ed.) Toxic Constituents of Plant Foodstuffs, Academic Press (New York), Chapter 2.

32. Rackis, J.J. 1972. In (A.K. Smith and S.J. Circle, eds.) Soybeans, Chemistry and Technology, Vol. 1, AVI Publ. Co. Inc. (Westport, CT), Chapter 6.

33. Booth, A.W., A.J. Robbins, W.E. Rebelin, F.D. (eds.) 1960. Effect of raw soybean meal and amino acids on pancreatic hypertrophy in rats. Proc. Soc. Exptl. Biol. Med. 104:681-683.

34. Bowman, D.E. 1944. Fractions derived from soybeans and navy beans which retard the tryptic digestion of casein. Proc. Soc. Exptl. Biol. Med. 57:139-140.

35. Eldridge, A.C., R.L. Anderson, W.J. Wolf. 1966. Polyacrylamide gel electrophoresis of soybean whey proteins and trypsin inhibitors. Arch. Biochem. Biophys. 115:495-504.

36. Frattali, V., R.F. Steiner. 1968. Soybean inhibitors. I. Separations and some properties of three inhibitors from commercial crude soybean trypsin inhibitor. Biochem. 7:521-530.

37. Kunitz, M. 1945. Crystallization of a soybean trypsin inhibitor from soybean. Science 101:668-669.

38. Birk, Y. 1961. Purification and some properties of a highly active inhibitor of trypsin and chymotrypsin from soybeans. Biochem. Biophys. Acta 54:378-381.

39. Yamamoto, M., T. Ikenaka. 1967. Studies on soybean trypsin inhibitors. I. Purification and characterization of two soybean trypsin inhibitors. J. Biochem. (Tokyo) 62:141-149.

40. Koide, T., T. Ikenada. 1973. Studies on soybean trypsin inhibitors: 3. Amino-acid sequence of the carboxyl-terminal region and the complete amino-acid sequence of soybean trypsin inhibitor (Kunitz). Eur. J. Biochem. 32:417-431.

41. Hymowitz, T. 1973. Electrophoretic analysis of SBTI-A$_2$ in the USDA soybean germplasm collection. Crop Sci. 13:420-421.

42. Hymowitz, T., J.H Orf, N. Kaizuma, H. Skorupska. 1978. Screening the USDA soybean germplasm collection for Kunitz trypsin inhibitor mutants. Soybean Genet. Newsl. 5:19-22.

43. Clark, R.W., D.W. Mies, T. Hymowitz. 1970. Distribution of a trypsin inhibitor variant in seed proteins of soybean varieties. Crop Sci. 10:486-487.

44. Hymowitz, T., D.W. Mies, C.J. Klebek. 1971. Frequency of a trypsin inhibitor variant in seed protein of four soybean populations. East Afr. Agr. For. J. 37:62-72.

45. Orf, J.H. 1976. Electrophoretic studies on seed proteins of Glycine max (L.) Merrill. M.S. Thesis, University of Illinois, Urbana.

46. Skorupska, H., T. Hymowitz. 1977. On the frequency distribution of alleles of two seed proteins in European soybean [Glycine max (L.) Merrill] germplasm: Implications on the origin of European soybean germplasm. Genetica Polonica 18:217-224.

47. Singh, L., C.M. Wilson, H.H. Hadley. 1969. Genetic differences in soybean trypsin inhibitors separated by disc electrophoresis. Crop Sci. 9:489-491.

48. Orf, J.H., T. Hymowitz. 1979. Inheritance of the absence of the Kunitz trypsin inhibitor in seed protein of soybeans. Crop Sci. 19:107-109.

49. Hymowitz, T., H.H. Hadley. 1972. Inheritance of a trypsin inhibitor variant in seed protein of soybeans. Crop Sci. 12:197-198.

50. Orf, J.H., T. Hymowitz. 1977. Inheritance of a second trypsin inhibitor variant in seed protein of soybeans. Crop Sci. 17:811-813.

51. Hwang, D.L., W.K. Yang, D.E. Foard, K.T. Davis Lin. 1978. Rapid release of protease inhibitors from soybeans. Immuno-chemical quantitative and parallels with lectins. Plant Physiol. 61:30-34.

52. Hildebrand, D.F., J.H. Orf, T. Hymowitz. 1980. Inheritance of an acid phosphatase and its linkage with the Kunitz trypsin inhibitor in seed protein of soybeans. Crop Sci. 20 (in press).

53. Jaffe, W.G. 1969. In: (I.E. Liener, ed.) Toxic Constituents of Plant Foodstuffs, Academic Press (New York) Chapter 3.

54. Weinhaus, O. 1909. Zur biochemic des phasins. Biochemische Zeitschrift 18:228-260.

55. Liener, I.E., M.J. Pallansch. 1952. Purification of a toxic substance from defatted soybean meal. J. Biol. Chem. 197:29-36.

56. Catsimpoolas, M., E.W. Meyer. 1969. Isolation of soybean hemagglutinin and demonstration of multiple forms by isoelectric focusing. Arch. Biochem. Biophys. 132:279-285.

57. Fountain, D.W., W. Yang. 1977. Isolectins from soybean (Glycine max) Biochem. Biophys. Acta 492:176-185.

58. Lis, H., C. Fridman, N. Sharon, E. Katchalski. 1966. Multiple hemagglutinins in soybean. Arch. Biochem. Biophys. 117:301-309.

59. Rackis, J.J., H.A. Sasame, R.L. Anderson, A.K. Smith.
    1959. Chromatography of soybean whey proteins on
    diethylaminoethylcellulose. J. Am. Chem. Soc.
    81:6265-6270.

60. Stead, R.H., H.J.H. DeMuelenaere, G.V. Quicke. 1966.
    Trypsin inhibitor, hemagglutination and intraperito-
    neal toxicity in extracts of Phaseolus vulgaris and
    Glycine max. Arch. Biochem. Biophys. 113:703-708.

61. Liener, I.E., J.E. Rose. 1953. Soyin, a toxic protein
    from the soybean. III. Immunochemical properties.
    Proc. Soc. Exptl. Biol. Med. 83:539-544.

62. Lotan, R., H.W. Siegelman, H. Lis, N. Sharon. 1974.
    Subunit structure of soybean agglutinin. J. Biol.
    Chem. 249:1219-1224.

63. Lotan, R., R. Cacan, M. Cacan, H. Debray, W.G. Carter,
    N. Sharon. 1975. On the presence of two types of
    subunit in soybean agglutinin. Fed. Europ. Biochem.
    Soc. Lett. 75:100-103.

64. Lis, H., B. Sela, L. Sachs, N. Sharon. 1970. Specific
    inhibition by N-acetyl-D-galactosamine of the
    interaction between soybean agglutinin and animal
    cell surfaces. Biochem. Biophys. Acta. 211:582-585.

65. Pull, S.P. 1978. An analysis of soybean lectin content
    in the seeds of 51 lines of Glycine max (L.) Merr.
    M.S. Thesis, University of Missouri, St. Louis.

66. Orf, J.H. 1979. Genetic and nutritional studies of seed
    lectin, Kunitz trypsin inhibitor, and other proteins
    of soybean [Glycine max (L.) Merrill]. Ph.D.
    Dissertation, University of Illinois, Urbana.

67. Pull, S.P., S.G. Pueppke, T. Hymowitz, J.H. Orf. 1978.
    Screening soybeans for lectin content. Soybean Genet.
    Newsl. 5:66-70.

68. Pull, S.P., S.G. Pueppke, T. Hymowitz, J.H. Orf. 1978.
    Soybean lines lacking the 120,000 dalton seed lectin.
    Science 200:1277-1279.

69. Orf, J.H., T. Hymowitz, S.P. Pull, S.G. Pueppke. 1978.
    Inheritance of a soybean seed lectin. Crop Sci. 18:
    899-900.

70. Liener, I.E. 1953. Soyin, a toxic protein from the soy-
    bean. I. Inhibition of rat growth. J. Nutr. 49:
    527-539.

71. Bahlool, B.B., E.L. Schmidt. 1974. Lectins: a possible
    basis for specificity in the Rhizobium-legume root
    nodule symbiosis. Science 185:269-271.

72. Bhuvaneswari, T.V., S.G. Pueppke, W.D. Bauer. 1977.
    Binding of soybean lectin to rhizobia. Plant Physiol.
    60:486-491.

73. Bahlool, B.B., E.L. Schmidt. 1976. Immunofluorescent
    polar tips of Rhizobium japonicum: possible site of
    attachment of lectin binding. J. Bacteriol. 125:
    1188-1194.

74. Kauss, H., C. Glaser. 1974. Carbohydrate binding pro-
    teins from plant cell walls and their possible
    involvement in extension growth. FEBS Letters 45:
    304-307.

75. Reporter, M., D. Raveed, G. Norris. 1975. Binding of
    Rhizobium japonicum to cultured soybean cell roots:
    morphological evidence. Plant Sci. Lett. 5:73-76.

76. Liener, I.E. 1974. Phytohemagglutinins: their nutri-
    tional significance. J. Agric. Food Chem. 22:17-23.

77. Janzen, D.H., H.B. Juster, I.E. Liener. 1976.
    Insecticidal action of the phytohemagglutinin in
    black beans on a bruchid beetle. Science 192:795-796.

78. Mirelman, D., E.E. Galun, N. Sharon, R. Lotan. 1975.
    Inhibition of fungal growth by wheat germ agglutinin.
    Nature 256:414-416.

79. Birk, Y., A. Gertler. 1961. Effect of mild chemical
    and enzymatic treatments of soybean meal and soybean
    trypsin inhibitors on their nutritive and biochemical
    properties. J. Nutr. 75:379-387.

80. Turner, R.H., I.E. Liener. 1975. The effect of the selective removal of hemagglutinins on the nutritive value of soybeans. J. Agric. Food Chem. 23:484-487.

81. Orf, J.H., T. Hymowitz. 1979. Soybean linkage test between Ti and Le seed proteins. Soybean Genet. Newsl. 6:32.

82. Kinsella, J.E. 1979. Functional properties of soy proteins. J. Am. Oil Chemists' Soc. 56:242-258.

83. Gertler, A., Y. Birk. 1965. Purification and characterization of β-amylase from soya beans. Biochem. J. 95: 621-627.

84. Greenwood, C.T., A.W. Macgregor, E.A. Milne. 1965. Starch degrading enzymes. II. The β-enzyme from soybeans; purification and properties. Carbohydrate Res. 1:229-241.

85. Peat, S., W.J. Whelan, S.J. Pirt. 1949. The amylolytic enzymes of soybean. Nature 164:499-500.

86. Birk, Y., M. Waldman. 1965. Amylolytic, trypsin-inhibiting, and urease-activity in three varieties of soybeans and in the soybean plant. Qualitias Plantarum et Materiae Vegetabiles 12:200-209.

87. Fukumoto, J., Y. Tsujisaka. 1954. Studies on soybean amylase. Purification and crystallization of the β-amylase of soybean. Kagaku to Kogyo, Osaka 28:282-287 (in Japanese).

88. Morita, Y., F. Yagi, S. Aibara, H. Yamashita. 1976. Chemical composition and properties of soybean β-amylase. J. Biochem. 79:591-603.

89. Gorman, M.B., Y.T. Kiang. 1977. Variety specific electrophoretic variants of four soybean enzymes. Crop Sci. 17:963-965.

90. Gorman, M.B., Y.T. Kiang. 1978. Models for the inheritance of several variant soybean electrophoretic zymograms. J. Heredity 69:255-258.

91. Hildebrand, D.F., T. Hymowitz. 1980. The Sp_1 locus in soybean codes for β-amylase. Crop Sci. 20: 165-168.

92. Larsen, A.L. 1967. Electrophoretic differences in seed proteins among varieties of soybean, Glycine max (L.) Merrill. Crop Sci. 7:311-313.

93. Larsen, A.L., B.E. Caldwell. 1968. Inheritance of certain proteins in soybean seed. Crop Sci. 8:474-476.

94. Orf, J.H., T. Hymowitz. 1976. The gene symbols Sp_1^a and Sp_1^b assigned to Larsen and Caldwell's seed protein bands A and B. Soybean Genet. Newsl. 3:27-28.

95. Hildebrand, D.F., T. Hymowitz. 1979. Inheritance of the lack of β-amylase activity in soybean seed. Agron. Abs., p. 63.

96. Hymowitz, T., N. Kaizuma, J.H. Orf, H. Skorupska. 1979. Screening the USDA soybean germplasm collection for Sp_1 variants. Soybean Genet. Newsl. 6:30-32.

97. Wilson, L.S., V.A. Birmingham, D.P. Moon, H.E. Snyder. 1978. Isolation and characterization of starch from mature soybeans. Cereal Chem. 55:661-670.

98. Yazdi-Samadi, B., R.W. Rinne, R.D. Seif. 1977. Components of developing soybean seeds: oil, protein, sugars, starch, organic acids, and amino acids. Agron. J. 69:481-486.

99. Dunn, G. 1974. A model for starch breakdown in higher plants. Phytochem. 13: 1341-1346.

100. Whittaker, R.H., P.P. Feeny. 1971. Allelochemics: chemical interactions between species. Science 171:767-770.

101. Ehrlich, P.R., P.H. Raven. 1967. Butterflies and plants. Sci. Amer. 216:104-113.

102. Freedland, W.J., D.H. Janzen. 1974. Strategies in herbivory by mammals: the role of plant secondary compounds. Amer. Natur. 108:269-289.

103. Seigler, D.S. 1977. Primary roles for secondary compounds. Biochem. Syst. & Ecol. 5:195-199.

Chapter Eight

CITRUS ESSENTIAL OILS: EFFECTS OF ABSCISSION CHEMICALS
AND EVALUATION OF FLAVORS AND AROMAS*

MANUEL G. MOSHONAS AND PHILIP E. SHAW

Citrus and Subtropical Products Laboratory
USDA
Winter Haven, FL

Introduction
Effects of chemicals used to control fruit abscission
Flavor effects of abscission chemicals
Compositional effects of abscission chemicals
Application of essential oil analyses

INTRODUCTION

Citrus oils are the most important by-products derived
from citrus fruits and are primarily used as flavoring,
scenting or masking agents for a wide variety of food,
beverage, pharmaceutical and perfumery products. Thus,
they contribute to the acceptance of many finished products
purchased by millions of consumers. Essential oils from
citrus have been sold commercially since 1582, when they
were listed in the Frankfurt am Main Statutes.

Citrus essential oils are obtained from small ductless
glands which are present in the flavedo or outer layer of
the peel. There has been a great deal of applied research
carried out on citrus essential oils, but we will focus on

*   Mention of a trademark name or proprietary product is
for identification only and does not constitute a guarantee
or warranty of the product by the U.S. Department of
Agriculture and does not imply its approval to the exclusion of other products which may also be suitable.

only two areas: (a) the effects of chemicals used to control fruit abscission necessary for mechanical harvesting; and (b) the evaluation of flavor and aroma compounds in oils from the five major citrus fruit: orange, grapefruit, mandarin, lemon and lime.

The largest single source of citrus oil is the 190 million or so boxes of oranges harvested and processed annually in Florida. The increase in the production of fruit available for processing and difficulties in retaining a stable work force have accelerated research to develop mechanical harvesting systems for citrus fruit. The most promising systems being tested for citrus require the prior application of abscission-inducing chemicals to the tree to loosen the fruit. These chemicals alter metabolic pathways within the fruit, causing it to loosen on the tree. They also affect the composition of the oil which in turn can affect flavor quality of products to which the oil is added.

The essential oil from each citrus species imparts the unique, fresh, fruity, complex aroma and flavor associated with that type of fruit. Cold-pressed oils are used in citrus juice concentrates, soft drinks, extracts, syrups, candies, jellies, baked goods, sherbets, perfume, soap, paper, paint thinners and other products, and are also used in combination for special aroma or flavor effects. Of the several different citrus fruit oils, cold-pressed orange oil is the most important commercially because of the large volume collected, even though it brings the smallest price per unit weight. Lime and lemon oils are the most valuable per unit weight, followed by tangerine, grapefruit and orange oils.

The chemical composition of citrus essential oils is important to growers, processors and product manufacturers in order to produce and maintain consistent, high quality consumer products. This knowledge establishes the contribution of specific compounds to flavor and aroma. It demonstrates the uniqueness of oils and the effects of changes in growing, harvesting or processing. This information also facilitates detecting decomposition, adulteration and cause of off-flavors, and would provide a basis for measuring and controlling quality of the oils.

EFFECTS OF CHEMICALS USED TO CONTROL FRUIT ABSCISSION

The most serious problem associated with development of a mechanical harvesting system for oranges has been the high force required to separate the orange from its stem. Simple direct mechanical shaking of trees, which had proven successful in the harvesting of other fruit, could not be used with oranges which have a fruit removal force as high as 22 lb. This compares with 6 lb for apples and 2 lb for cherries. Researchers thus began the study of chemicals which might reduce the removal force, making mechanical harvesting of citrus feasible.

In 1969 Cooper et al. reported on the potency of 3-[2-(3,5-dimethyl-2-oxocylohexyl)]-glutarimide (cyloheximide or Acti Aid) to promote abscission of citrus fruit.[6] Cooper and Henry[5] found that cycloheximide could be successfully used for early- and mid-season oranges. It was not satisfactory for late-season Valencia oranges because of excessive droppage of young fruit which the tree carries for the next season's crop. This problem has been overcome by use of two other chemicals. These chemicals, which aid abscission of mature oranges but do not damage or cause removal of the young fruit, are 5-chloro-3-methyl-4-nitro-1H-pyrozol (Release), reported by Wilson[28] and Kenny et al.[10], and glyoxal dioxime (Pik-Off), reported by Wilcox et al.[27] All three compounds act by damaging the peel, causing the release of "wound" ethylene, which promotes abscission.[5] A fourth compound found to aid abscission of citrus, 2-chloro-ethylphosphonic acid (ethephon), does not injure the rind but, to date, has been of limited use for loosening fruit.

There is a great need to examine the effect of these chemicals on quality, flavor and composition of essential peel oils and juice from oranges treated with these chemicals, if the future production of high quality products is to be assured.

FLAVOR EFFECTS OF ABSCISSION CHEMICALS

Flavor evaluations of peel oil and single strength juice from abscission treated oranges indicated they were different from similar products from untreated control oranges. Triangle and paired comparison tests, as described by Boggs and Hanson,[4] were used for these evaluations.

Results of flavor evaluation with peel oils are shown in Table 1. Each juice sample was prepared by the addition of 0.020 wt % oil by volume (w/v). In all but two evaluations the trained taste panel distinguished a difference between control and experimental samples at a confidence level of 95% or greater. Without exception when a difference was detected, the panels preferred the control sample.[16] Many panelists noted an "overripe" flavor in juice containing oil from treated fruit.

In one of the two tests showing no difference or preference, ethephon was the abscission agent that had been used. Ethephon acts by decomposing chemically to release ethylene rather than by injuring the rind to cause release of wound ethylene. Thus, ethephon might be expected to cause less metabolic and, therefore, less flavor changes than rind-injuring chemicals. However, ethephon has limited use as an abscission agent for citrus because it causes excessive leaf drop. The other test that showed no differences or preferences involved well-matured Valencia oranges sprayed with 250 ppm Release. It had since been determined that 400 ppm or more Release may be needed for adequate loosening of well-matured Valencias. The amount used for that study may not have been sufficient for full abscission effects and thus the flavor was not affected as much as in the Valencia samples harvested earlier that season.

Flavor evaluation of juice from oranges treated with abscission chemicals is also important since small amounts of essential oil enter the juice during processing. In addition, ongoing internal metabolic processes probably are affected by abscission chemicals. These chemicals may cause an accelerated maturing of the fruit or have other effects on flavor quality. Table 2 shows the results of flavor evaluations comparing juice from early- (Hamlin) and mid- (Pineapple) season oranges, and Table 3 shows results for late-season (Valencia) oranges treated with abscission chemicals compared with control oranges. Juice from Hamlin, Pineapple and Valencia experimental oranges treated with any of the abscission agents were distinguished from nontreated samples by the taste panel. Many panel members

TABLE 1

FLAVOR EVALUATION OF COLD-PRESSED OILS FROM EARLY-, MID-,
AND LATE-SEASON ORANGES TREATED WITH OR WITHOUT
ABSCISSION CHEMICALS

Cultivar maturity	Abscission agent	Concentration of agent, ppm	Days on tree after spray	Flavor evaluation: confidence level	
				Difference exptl vs control	Preference for control
Hamlin (early)					
Barely[a]	Acti Aid	20	7	0.01	0.05
Well	Acti Aid	20	7	0.01	0.05
Pineapple (mid)					
Barely	Acti Aid	20	6	0.001	0.05
Well	Acti Aid	20	6	0.001	0.01
Barely	Acti Aid	10	6	0.01	0.01
Well	Acti Aid	10	6	0.01	0.01
Barely	Ethephon	250	6	N.S.[b]	N.S.
Valencia (late)					
Barely	Acti Aid	20	7	0.001	0.01
Well	Acti Aid	20	7	0.01	0.01
Barely	Release	250	5	0.001	0.01
Well	Release	250	4	N.S.	N.S.
Barely	Pik-Off	300	5	0.05	0.05
Well	Pik-Off	300	4	0.01	0.01

[a] Time when fruit first reaches legal maturity

[b] Not significant at 0.05 confidence level or greater

TABLE 2

FLAVOR EVALUATION OF JUICE FROM EARLY- AND MID-SEASON ORANGES
TREATED WITH OR WITHOUT (CONTROL) ABSCISSION CHEMICALS

Sample no.	Maturity	Acid ratio	Abscission agent	Spray of agent (ppm)	Days on tree after spray	Flavor evaluation: confidence level	
						Difference exptl vs control	Preference for control
				Hamlin oranges (early)			
1	Barely[a]	10.45	Acti Aid	20	7	0.05	N.S.[b]
2	Well	12.86	Acti Aid	20	7	0.001	0.01
				Pineapple oranges (mid)			
3	Barely	9.87	Acti Aid	20	6	0.001[c]	0.01[c]
4	Well	17.13	Acti Aid	20	6	0.001	0.01
5	Barely	12.22	Acti Aid	10	6	0.001[c]	0.01[d]
6	Well	13.95	Acti Aid	10	6	0.001	0.01
7	Barely	13.22	Ethephon	250	6	0.01	N.S.

a Time when fruit first reaches legal maturity
b Not significant at 0.05 confidence level or greater
c For both single-strength juice and concentrate
d Not significant for concentrate

TABLE 3

FLAVOR EVALUATION OF JUICE FROM VALENCIA ORANGES
TREATED WITH OR WITHOUT ABSCISSION CHEMICALS

Sample no.	Maturity	°Brix/ acid ratio	Abscission agent	Spray concentration of agent (ppm)	Days on tree after spray	Flavor evaluation: confidence level	
						Difference exptl vs control	Preference for control
1	Barely[a]	13.77	Acti Aid	20	7	0.001	0.01
2	Well	15.94	Acti Aid	20	7	0.01	0.01
3	Barely	10.44	Release	250	5	0.001	0.01
4	Well	12.02	Release	250	4	0.001	0.01
5	Barely	9.70	Release	150	3	0.001	_[b]
6	Barely	9.56	Release	150	5	0.001	0.01
7	Barely	10.61	Pik-Off	300	5	0.001	0.01
8	Well	11.61	Pik-Off	300	4	0.01	N.S.[c]
9	Barely	10.75	Pik-Off	300	3	N.S.	_[b]
10	Barely	10.01	Pik-Off	300	6	0.001	0.01

[a] Time when fruit first reaches legal maturity
[b] Tests not run
[c] Not significant at 0.05 confidence level or greater

again indicated juice from treated oranges had an "overripe" flavor which was considered adverse. In most tests
the panel preferred the control juice. No preference was
established in a few comparisons involving either very
acidic or very sweet samples. Probably, detection of
differences was influenced or masked by the extreme sour or
sweet conditions. In no test was the experimental juice
preferred by the panel.

COMPOSITIONAL EFFECTS OF ABSCISSION CHEMICALS

    Analysis of nineteen essential oils from Hamlin,
Pineapple and Valencia oranges treated with rind-injuring
abscission agents showed they affect a particular metabolic
pathway within the orange. Regardless of cultivar,
rootstock, maturity, fertilizer or irrigation practices,
use of the chemicals resulted in production of the same
group of phenolic ethers. None of the phenolic ethers were
found in any of the nineteen corresponding control
essential oils from untreated oranges. Essential oil from
oranges treated with the non-injuring chemical, ethephon,
did not indicate presence of the phenolic ethers.

    The six major phenolic ethers (which have never been
reported as natural components of citrus fruit) are:
eugenol, methyleugenol, cis-methylisoeugenol, trans-
methylisoeugenol, elemicin and isoelemicin. Figure 1 shows
the structure of each in a scheme of pathways by which they
might be formed. The proposed pathways are based on findings of Tressl and Drawert.[26] They reported that banana
discs converted [1-^{14}C] caffeic acid into labeled eugenol,
methyleugenol and elemicin. The phenolic acids and enzymes
involved in these reactions have been reported as present
in citrus fruit.[20]

    Table 4 shows flavor thresholds of four of the phenolic
ethers and estimated amounts of all six phenolic ethers
found in single-strength orange juice. The most potent was
eugenol, with a flavor threshold of 22 ppb, followed by
methyleugenol, 1.25 ppm, elemicin, 22 ppm, and trans-
methylisoeugenol, 35 ppm. Although these compounds were
present at below the flavor threshold in orange juice, they
are present in concentrations about equal to or higher than
threshold levels in other foods that an individual could
consume. Arctander[3] reported that eugenol, at 10 to over

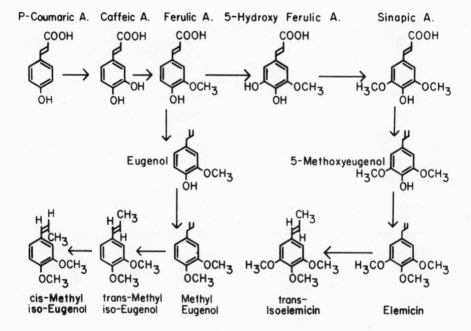

Figure 1. Possible pathways for formation of phenolic
          ethers in orange oils

TABLE 4
FLAVOR THRESHOLDS AND ESTIMATED AMOUNTS OF NEW COMPOUNDS
PRESENT IN JUICE FROM FRUIT TREATED WITH
ABSCISSION CHEMICALS

Compound	Flavor threshold	Amount present ppb
Eugenol	22	21
Methyleugenol	1250	42
Elemicin	22000	10
trans-Methylisoeugenol	35000	21
cis-Methylisoeugenol	ND[a]	4.2
trans-Isoelemicin	ND	31.5

[a] Not determined

100 ppm, is used to impart a clove-like note to combination spice flavors and to modify complex flavors such as those of nuts, mint and various fruits; that trans-methyliso-eugenol, at 5 to 100 ppm is used as a flavor component of spice blends, vanilla imitations and chocolate bases; and that methyleugenol, at 5 to 15 ppm is used as a flavor component of juice blends and baked goods. Flavor thresholds for cis-methylisoeugenol and isoelemicin were not determined because information regarding safe levels for human consumption was not available. The additive flavor contribution of three of the phenolic ethers was investigated by comparison of control orange juice with juice to which was added eugenol, trans-methylisoeugenol and methyleugenol at half their threshold levels. The taste panel distinguished the juice with the phenolic ethers at the 99.9% confidence level. Although none of the ethers was present at levels above their flavor threshold, the flavor effects apparently were additive and the combined effect was sufficient to account for off-flavor of processed single-strength orange juice reported in 1976.[18] Eugenol, in particular, was present in orange juice at about its flavor threshold (21 ppb). These phenolic ethers contributed to orange juice an overripe flavor note similar to that discerned in experimental juices by the taste panel.

The concentration of each phenolic ether in orange oil from abscission treated oranges was estimated from gas chromatographic peak areas. These oils were estimated to contain 120 ppm eugenol, 240 ppm methyleugenol, 180 ppm isoelemicin, 120 ppm trans-methylisoeugenol, 60 ppm elemicin, and 24 ppm cis-methylisoeugenol.[17] Increased amounts of phenolic ethers were found when oranges had been treated with higher concentrations of abscission chemicals.

Current processing practices tend to minimize flavor effects that the abscission chemicals may have on citrus products. In the first place, flavor differences noted by trained panelists may not be significant to the untrained consumer. Also, in the normal processes for citrus products, juice from fruit treated with abscission chemicals would be mixed with much larger volumes of juice from untreated fruit (since, currently, only a small fraction of fruit is mechanically harvested). Thus, the adverse flavor effect would be reduced by dilution. In addition, preparation of orange juice concentrate, for which most juice is used, tends to reduce, but not eliminate, flavor effects.[18]

However, as mechanical harvesting of citrus becomes more widely adopted, the factors that currently minimize adverse flavor effects will diminish. Thus, it will become increasingly important to carefully control the concentration of abscission agents and to minimize the time necessary for fruit to loosen after chemical treatment. A new generation of abscission chemicals has been developed that are combinations of those reported above. They loosen fruit effectively at a lower concentration than needed for the individual chemicals. Preliminary qualitative results have shown that the mixtures also induce formation of the same phenolic ethers. However, the levels may be low enough so that flavor effects are minimal.

APPLICATION OF ESSENTIAL OIL ANALYSES

Several types of flavor fractions produced from citrus oils, juices and peel are used by the citrus industry and other segments of the food industry in a variety of food products. Some of these are used also in cosmetics and household and industrial cleansers because of the almost universal appeal of citrus flavors and aromas, especially lemon and orange.

In Table 5 are listed the five major citrus fruit from which citrus flavor fractions are isolated and the major types of flavor fractions produced commercially. Also shown are sources of those fractions, principal uses and the fruit from which these fractions are produced commercially. Several of the flavor fractions listed have been produced and studied experimentally, but have not been produced commercially.

The fraction from citrus (except for lime) that is used the most widely as a flavor ingredient is cold-pressed oil. This is the essential oil expressed from the peel either simultaneously with juice extraction or immediately following juice extraction. A thorough review of citrus peel essential oil constituents has recently been published.[21] With lime, the major oil of commerce is a distilled oil produced from the Mexican lime. In this process, the whole fruit is crushed to produce a slurry; the peel oil thus released from the oil glands is efficiently removed by distillation. When separated by centrifugation without

distillation the oil is milder flavored and more costly.
In Mexico, the world's major producer of lime oil, about
90% of the lime oil is collected by the distillation
method.[9] Contact between the acidic juice and the peel
oil during the extraction-distillation process creates
artifacts that appear in the distilled oil and give it a
harsh, "reverted" flavor not typical of cold-pressed lime
peel oil. However, it is this distilled oil that is most
widely used in lime-flavored products. Recently, a cold-
pressed Mexican lime oil has been produced by a technique
that involves rasping the whole fruit peel and washing the
oil from the peel.[9] The flavor of this cold-pressed oil
and that of cold-pressed Persian lime oil is milder than
the traditional flavor of distilled lime oil when used in
limeades and other lime-flavored drinks. Reasons for the
relatively harsh flavor of distilled lime oil will be
discussed below.

     Several solutions to practical problems have resulted
from the many reported analyses of cold-pressed citrus
oils. In Table 6 are listed the individual components
believed important to the flavor of each major citrus
cultivar and some practical applications. Aldehydes have
been known for many years to be important to the characteristic flavors of orange, grapefruit and lemon oils.
Octanal, citral (a mixture of the two cis-trans isomers,
neral and geranial) and acetaldehyde were shown by Ahmed et
al.[2] to be important contributors to orange flavor.
Octanal is present in orange oil in much greater quantity
than either citral or acetaldehyde,[23] whereas acetaldehyde
is the major aldehyde in the juice.[22] The relative importance of the various aldehydes to orange flavor has not
been thoroughly assessed.

     (+)-Limonene is the main component of orange oil and of
other citrus oils. It is present in orange juice at a
level 400 times its taste threshold in water and, thus, is
an important contributor to the characteristic aroma and
flavor of the oil.[1] If the concentration of limonene is
reduced, as in concentrated oils, the flavor and strength
of the oil are altered. However, concentrated oil will
have less flavoring strength than would be expected based
on the strength of the original oil, partly because of the
decreased limonene content and partly because of loss of
other components in the concentration process.

TABLE 5
CITRUS FLAVOR FRACTIONS AND THEIR PRINCIPAL USES

Flavor Fraction	Source	Orange	Grapefruit	Mandarin	Lemon	Lime	Principal Uses
Cold-pressed oil	Expressed from peel	Com.[a]	Com.	Com.	Com.	Com.	Juice concentrates, synthetic beverages, candies, ice cream
Essence oil	Distilled from juice	Com.	Com.	Exp.[a]	Exp.	Exp.	Juice concentrates synthetic beverages
Aqueous essence	Distilled from juice	Com.	Com.	Exp.	Exp.	Exp.	Juice concentrates, single-strength juices
Conc. essence	Aqueous essence concd. by distillation	Exp.	N[a]	N	N	N	Juice concentrates
Aroma oil (stripper oil)	Distilled from centrifuge effluent or peel slurry	Com.	Com.	Com.	Com.	Exp.	Cleansers, source of (+)-limonene

[a] Com. = commercial product; Exp. = experimentally prepared and evaluated; N = not prepared.

TABLE 6
IMPORTANT FLAVOR COMPONENTS OF EACH MAJOR CITRUS

Cultivar	Flavor Component	Ref.	Practical Applications
Orange	Octanal Ethyl butyrate Citral Acetaldehyde	2	Total aldehydes used as quality index Esters indicated as quality index
	(+)-Limonene		Needed as carrier; concentrated oils have different flavor
	α- and β-Sinensals	7	
Grapefruit	Nootkatone	12	Used as measure of quality in synthetic drinks
	Decanal	13	Total aldehydes used as quality index
	Geranyl acetate Neryl acetate Octyl acetate Carvyl acetate Citronellyl acetate 1,8-p-Menthadien-9-yl-acetate		Nootkatone is not always present in high quality oils and acetate esters are suggested as important contributors to flavor
Mandarin	Dimethyl anthranilate Thymol Thymol methyl ether α-Sinensal	11 15	All are claimed to be important to mandarin flavor, but no firm evidence to support claims
Lemon	Citral (neral + geranial) Neryl acetate Geranyl acetate	29	Aldehyde content used as quality factor
Lime (distilled)	1,8-Cineole p-Cymeme α-Terpineol Citral	25	Use in synthetic lime oil for lime flavoring of candies, carbonated beverages

One other component of orange oil that contributes a
strong orange flavor and aroma note is sinensal, found in
orange oil as a mixture of α- and β-isomers. Because
sinensal is unstable in the pure form, its contribution to
orange flavor has been difficult to assess. Nevertheless,
it had been reported to contribute significantly to orange
flavor.[29]   Ohloff[19] stated that sinensal is important to
orange flavor. He reported that the α-isomer has an orange-
like aroma, but that the β-isomer has a "strong metallic-
fishy undertone" and an unpleasant aroma at high concentrations. To date no practical use has been made of the
presence of sinensal in orange oil, nor have sinensal-
enriched fractions been used as flavor fractions or flavor
enhancers. A fraction from orange oil enriched in sinen-
sals is much more stable than pure sinensal and, thus, may
have potential as a citrus flavor fraction.

Nootkatone is a sesquiterpene ketone that is a principal flavor-influencing compound in grapefruit oil (Table
6). As a result of its identification by MacLeod and
Buigues[12] in grapefruit oil and its demonstration as an
important contributor to grapefruit flavor, it has been
synthesized commercially and is used for flavoring carbonated grapefruit-flavored beverages. The nootkatone content
in grapefruit oil is one measure of oil quality. Some
full-flavored grapefruit oils have been found with almost
no nootkatone, however, indicating other constituents are
also important in grapefruit flavor. Several terpene
acetate esters were identified in grapefruit oil by
Moshonas,[13] who suggested their importance to grapefruit
flavor. So far, nootkatone is the only grapefruit oil
constituent that has been shown to contribute significantly
to a distinct grapefruit flavor.

Dancy tangerine oil is from a type of mandarin fruit
that has a uniquely desirable flavor, and this oil brings a
premium in flavor markets. The term "mandarin" is a
general term referring to loose skin citrus cultivars and
most mandarins are hybrids between tangerine and some other
fruit.[23]   Little definitive information exists on the
constituents of mandarin oil responsible for the characteristic flavor, despite the many detailed analyses on the
oils and claims of importance to flavor for certain constituents. The compounds which are probably most significant
to mandarin flavor are listed in Table 6. Kugler and
Kovats[11] indicated dimethyl anthranilate and thymol were

important contributors to mandarin flavor. Thymol methyl
ether and α-sinensal in tangerine peel oil were identified,
quantitated and indicated to possibly be important to mandarin flavor.[15] No taste panel studies or other definitive
evidence are available to designate any particular compounds as important to mandarin flavor.

   Lemon oil is the most valuable cold-pressed citrus oil
collected in large quantities. Cold-pressed lime oil is
slightly more valuable per unit weight than lemon oil, but
is produced in much smaller quantity. Lemon oil is widely
used in cola-flavored and lemon/lime-flavored carbonated
beverages, as well as in alcoholic drink mixes, lemonade,
candies, cake mixes, cosmetics, cleansers and furniture
polish. The major flavor influencing compound in lemon oil
has long been known to be citral and the quality of lemon
oil is highly dependent on citral content. The best oils
contain 2-4% citral and the presence of more than 5%
suggests that the oil may have been adulterated.[29] Lemon
oils have been reported, however, that naturally contain up
to 13% citral.[8] Other components of lemon oil believed
important to desirable full-bodied flavor are neryl acetate
and geranyl acetate. They have not been shown to be essential to lemon flavor, however, and no limits have been set
on their presence in high quality lemon oils. The presence
of appreciable quantities of the lime oil constituents 1,8-
cineole, α-terpineol and p-cymene would be considered
detrimental in lemon oil, since they contribute a harsh
flavor and are indicative of abuse during collection or
storage of the oil.

   Distilled lime oil, like lemon oil, is widely used in
carbonated cola-flavored and lemon/lime-flavored drinks as
well as in candies and other foods. Synthetic lime oils
are available that resemble the strong flavor of distilled
lime oil. Synthetic oils for the other common citrus
flavors are less-widely used since the complex, delicate
flavor characteristic of other citrus oils is harder to
duplicate synthetically. Distilled lime oil has a relatively low citral content of only 0.3%, compared to 3-5% for
cold-pressed lime oil.[23] The compounds important to
distilled lime flavor are shown in Table 6. In addition to
citral, the presence of 1,8-cineole, p-cymene and
α-terpineol are necessary to impart a typical lime flavor.
The latter three compounds are produced by acid-catalyzed
degradation of the lime peel oil components β-pinene,

citral and γ-terpinene during the distillation process to
isolate the oil, when the acidic juice is in contact with
the oil. In distilled lime oil the levels of β-pinene,
citral and γ-terpinene are one-tenth the levels in cold-
pressed lime oil.[9,21]

In all of these citrus oils, further work is needed on
the precise mixture of components necessary for flavor in
the most desirable oils. There is also need to develop
rapid, simple analytical techniques to monitor the components essential to high quality oils. Another area in
need of much research is the study of storage degradation
of oils alone, or in the products in which they are used.

Two other flavor fractions of increasing commercial
importance, especially in orange products, are essence oil
and aqueous essence (Table 5). These two fractions are
part of, and usually separated from, the condensed
distillate from the first or second stage of the evaporator
used in making concentrated citrus juices. They comprise
an oil and aqueous phase that have been equilibrated during
distillation and condenstion and are separated after
condensation. The oil phase (essence oil) is enriched in
the more oil-soluble volatile components and the aqueous
phase (aqueous essence) in the more water-soluble
components. Both phases may be added back to the frozen
concentrated juice to provide "fresh" flavor notes which
were removed during evaporation. Some of these flavor
notes are not provided by the cold-pressed peel oil which
is traditionally added to provide more flavor to frozen
concentrated citrus juices. To increase the stability of
essence oil, it is usually blended with cold-pressed peel
oil which contains natural antioxidants not present in
essence oil.

Analytical studies on orange aqueous essence have been
extensive (reviewed by Shaw[22]). The major organic component of aqueous essence is ethanol, which contributes
little to the flavor of essence. Its main function is
perhaps as a carrier for the minor flavor components.
Acetaldehyde is the constituent present in largest quantity
that has been shown to make a positive contribution to
orange flavor.[2] Other constituents, present in trace
amounts, shown to be important to orange flavor, are
octanal, citral, several volatile esters, (ethyl butyrate,
methyl butyrate, ethyl acetate, and ethyl propionate, which

contribute a fruit note) and 2-pentenal, which contributes an astringent, fruity note at low concentration.[14] 1-Penten-3-one, which has a low flavor threshold, may contribute a positive or negative note, depending on concentration.[14,22]

There is a need within the citrus industry for objective methods to measure essence strength and quality so that blends of uniform quality might be produced. Many efforts have been made to correlate individual components or classes of components (aldehydes, esters, alcohols) in aqueous orange essence with flavor quality and strength. These efforts have been partly successful, but flavor evaluation is still the final test of strength and quality for each batch of essence.[22]

To date, all components reported present in orange essence oil have been identified in either cold-pressed orange oil or aqueous orange essence. Since essence oil contains little of the powerfully flavored volatile components, such as acetaldehyde and 1-penten-3-one, and since it is usually blended with cold-pressed oil, the need to standardize essence oil is not as critical as for aqueous essence. In some citrus plants small amounts of aqueous essence and/or essence oil are used in citrus products other than frozen concentrated juice, such as single strength juice packed in glass.

Few studies have been reported on storage retention or stability of aqueous essence and essence oil in products other than frozen concentrated juice. However, the extensive background of analytical work reported for these flavor fractions makes such studies feasible in the near future.

Essences and essence oils from citrus other than orange have been much less widely used to flavor products. Analytical studies have shown that grapefruit essence and essence oil contain the principal flavor influencing compound, nootkatone, and that lemon and lime aqueous essences contain the principal flavor influencing compound, citral.

A concentrated aqueous orange essence has been prepared commercially with potential for use in distant markets where shipping costs are large. This concentrate contains only 25% water as compared with 85% water content in normal aqueous essence. Analytical studies comparing composition of commercial aqueous essence with concentrated essence show no significant differences with regard to flavor components.[24]

"Stripper oils" are recovered from various waste streams in a citrus plant, and their aromas vary from a pleasant, dilute, orange aroma to a harsh, burnt-caramel aroma, depending on feed stock and method of collection. This is the last flavor fraction listed in Table 5. Most is sold for its (+)-limonene content (ca. 97%), but some is used in industrial-type cleaners where the citrus aroma may mask other more harsh aromas. Stripper oil usually sells for about the price of (+)-limonene but its recovery aids in waste disposal. The presence of oils in citrus waste streams can disrupt the natural balance of organisms necessary for adequate waste treatment.

In conclusion, compositional studies of citrus flavor fractions have helped solve several practical problems for the citrus, flavor and food industries and suggest solutions to other problems. These studies have led to standards of identity for citrus oils and suggest standards of identity and means for objective blending of aqueous essences to achieve a uniform product. Analytical studies on flavor fractions from waste streams have suggested use of these fractions as additional by-products to increase return on money invested. The extensive background of information now available on composition of citrus oils and essences will help develop future opportunities for their use. Solutions to problems encountered in the use of these flavor fractions will be easier than in the past.

REFERENCES

1. Ahmed, E.M., R.A. Dennison, R.H. Dougherty, P.E. Shaw. 1978. Flavor and odor thresholds in water of selected orange juice components. J. Agric. Food Chem. 26:187.

2. Ahmed, E.M., R.A. Dennison, P.E. Shaw. 1978. Effect of selected oil and essence volatile components in flavor quality of pumpout orange juice. J. Agric. Food Chem. 26:368.

3. Arctander, S. 1969. Perfume and Flavor Chemicals, Vols. I α II. Steffen Arctander (Montclair, NJ).

4. Boggs, M.M., H.L. Hanson. 1949. Advances Food Research, Vol. 2. Academic Press (New York, NY), p. 222.

5. Cooper, W.C., W.H. Henry. 1973. In: Shedding of Plant Parts, Academic Press (New York, NY), p. 475.

6. Cooper, W.C., W.H. Henry, G.K. Rasmussen, C.J. Hearn. 1969. Cycloheximide: an effective abscission chemical for oranges in Florida. Proc. Fla. State Hortic. Soc. 82:99.

7. Flath, R.A., R.E. Lundin, R. Teranishi. 1966. The structure of β-sinensal. Tetrahedron Lett. 295.

8. Gunther, H. 1968. Investigations on lemon oils by gas chromatography and infrared spectroscopy. Dtsch. Lebensm. Rudsch. 64:104.

9. Haro-Guzman, L., R. Huet. 1970. The essential oil of lime in Mexico. Fruits 25:887.

10. Kenney, D.S., R.K. Clark, W.C. Wilson. 1974. ABG-3030: An abscission chemical for processing oranges: biological activity. Proc. Fla. State Hortic. Soc. 87:34.

11. Kugler, E., E. Kovats. 1963. Information on mandarin peel oil. Helv. Chim. Acta 46:1480.

12. MacLeod, W.D., N.M. Buigues. 1964. Sesquiterpenes. I. Nootkatone, a new grapefruit flavor constituent. J. Food Sci. 29:565.

13. Moshonas, M.G. 1971. Analysis of carbonyl flavor constituents from grapefruit oil. J. Agric. Food Chem. 19:769.

14. Moshonas, M.G., P.E. Shaw. 1973. Some newly found orange essence constituents including trans-2-pentenal. J. Food Sci. 38:360.

15. Moshonas, M.G., P.E. Shaw. 1974. Quantitative and qualitative analysis of tangerine peel oil. J. Agric. Food Chem. 22:282.

16. Moshonas, M.G., P.E. Shaw. 1977. Effects of abscission agents on composition and flavor of cold-pressed orange peel oil. J. Agric. Food Chem. 25:1151.

17. Moshonas, M.G., P.E. Shaw. 1978. Compounds new to orange oil from fruit treated with abscission chemicals. J. Agric. Food Chem. 26:1288.

18. Moshonas, M.G., P.E. Shaw and D.A. Sims. 1976. Abscission agents effects on orange juice flavor. J. Food Sci. 41:809.

19. Ohloff, G. 1978. Recent developments in the field of naturally-occurring aroma components. In: Progress in the Chemistry of Organic Natural Products, Vol. 35, p. 431.

20. Raymond, W.R., V.P. Maier. 1977. Chalcone cyclase and flavonoid biosynthesis in grapefruit. Phytochemistry 16:1535.

21. Shaw, P.E. 1977. Essential oils. In: (Nagy, S., P.E. Shaw and M.K. Veldhuis, eds.) Citrus Science and Technology, Vol. 1. Avi Publ. Co. (Westport, CT), p. 427.

22. Shaw, P.E. 1977. Aqueous essences. In: (Nagy, S., P.E. Shaw and M.K. Veldhuis, eds.) Citrus Science and Technology, Vol. 1. Avi Publ. Co. (Westport, CT), p. 463.

23. Shaw, P.E. 1979. Review of quantitative analyses of citrus essential oils. J. Agric. Food Chem. 27:246.

24. Shaw, P.E., M.G. Moshonas. 1974. Analysis of concentrated orange essence and comparison with known essence composition. Proc. Fla. State Hortic. Soc. 87:305.

25. Slater, C.A., W.T. Watkins. 1964. Citrus essential oils. IV. Chemical transformations of lime oil. J. Sci. Food Agric. 12:732.

26. Tressl, R., F. Drawart. 1973. Biogenesis of banana
    volatiles. J. Agric. Food Chem. 21:560.

27. Wilcox, M., J.B. Taylor, W.C. Wilson, I.Y. Chen. 1974.
    Chemical abscission of 'Valencia' oranges by glyoxime
    (CGA-22911). Proc. Fla. State Hortic. Soc. 87:22.

28. Wilson, W.C. 1973. A comparison of cycloheximide with a
    new abscission chemical. Proc. Fla. State Hortic.
    Soc. 86:56.

29. Ziegler, E. 1971. The examination of citrus oils.
    Flavour Ind. 647.

INDEX

Tumor inhibitors (cont'd)
  newer varieties of, 30-35
Turnips, 143
Tylocrebrine, 28, 41
Tylophora spp., 28
Tylophora crebriflora, 28
Tylophorine, 28
Tylophorinine, 28

United States
  crop values and produc-
        tion in, 98-99
  domestic corn use in, 100
USDA, see Agriculture
        Department, U.S.

Valencia oranges, flavor
        evaluation of
        juice from, 184,
        187, see also
        Orange juice;
        Oranges
Vegetable oils, fatty acid
        composition of,
        121
Vegetable tanning material,
        9
Vernolepin, 29
Vicia faba, 67
Vinblastine, 24, 41
Vinca alkaloids, 41
Vinca rosea, 24
Vincristine, 24, 41
Vitis labrusca, 62
Vitis vinifera, 62

Walker carcinosarcoma-256
        26-27
Water-treatment compounds,
        9
Wax myrtle, 64
Waxy corn, 104-105
Weeds, 139
Welensali, 56
Wheat crop, size of, 97-98
Whole-plant oils, 2
  complex composition of, 6-7
  as industrial feedstocks, 6
  industrial utilization of, 10
  lipid classes in, 8
  plant sources of, 8
  separation costs for, 6-9
Wild species, botanochemical
        crop specifications
        and, 5
Wine
  aging of, 63
  as carcinogen, 61-63, 65
Wistar Institute, 68
"Wound" ethylene, release of,
        183

Yellow dock, 66

Zanthexylum spp., 31
Zein, of corn, 109